THE
INTERNATIONAL SERIES
OF
MONOGRAPHS ON PHYSICS

GENERAL EDITORS

J. BIRMAN   S. F. EDWARDS
C. H. LLEWELLYN SMITH   M. REES

Mechanisms of Conventional and High T_c Superconductivity

VLADIMIR Z. KRESIN

Lawrence Berkeley Laboratory
University of California at Berkeley

HANS MORAWITZ

IBM Almaden Research Center

STUART A. WOLF

Naval Research Laboratory

New York   Oxford
OXFORD UNIVERSITY PRESS
1993

ʋ5118967

Oxford University Press

Oxford   New York   Toronto
Delhi   Bombay   Calcutta   Madras   Karachi
Kuala Lumpur   Singapore   Hong Kong   Tokyo
Nairobi   Dar es Salaam   Cape Town
Melbourne   Auckland   Madrid

and associated companies in
Berlin   Ibadan

Copyright © 1993 by Oxford University Press, Inc.

Published by Oxford University Press, Inc.,
200 Madison Avenue, New York, New York 10016

Oxford is a registered trademark of Oxford University Press

Library of Congress Cataloging-in-Publication Data
Kresin, Vladimir Z.
Mechanisms of conventional and high T_c superconductivity /
Vladimir Z. Kresin, Hans Morawitz, Stuart A. Wolf.
p.   cm. — (The International series of monographs
on physics ; 84)
Includes bibliographical references and index.
ISBN 0-19-505613-2
1. Superconductivity.   2. High temperature superconductivity.
I. Morawitz, H.   II. Wolf, Stuart A.   III. Title.   IV. Series:
International series of monographs on physics (Oxford, England) ;
84.
QC611.92.K75   1993
537.6'23—dc20   92-40048

9 8 7 6 5 4 3 2 1
Printed in the United States of America
on acid-free paper

This book is dedicated to the memory of

Boris Geilikman
Leonard Schiff
Bernard Serin

PREFACE

This book presents a detailed review of various mechanisms of super-conductivity that have been proposed to provide high transition temperatures. These include phononic, magnetic, and electronic models. It contains the theoretical description as well as an analysis of crucial experiments. In addition, we carefully treat the phenomenon of induced superconductivity that manifests itself in both conventional and high T_c superconductors. The discovery of high T_c in the cuprates has stimulated intensive theoretical development. This book describes these advances and their links with the existing microscopic theory.

A variety of superconducting systems are described in the book, but the main focus is on the high transition temperature cuprates. A unified description of superconductivity in these materials is also presented.

We hope that physicists, chemists, materials scientists and graduate students interested in the theoretical development of superconductivity will benefit from this book.

V. Z. K.

H. M.

S. A. W.

ACKNOWLEDGMENTS

Over the last several years we have benefited from many discussions that had an impact on our thinking and are reflected in this book. In particular we are indebted to J. Ashkenazi, J. Bardeen, I. Bozovic, J. Cohn, G. Deutscher, G. Eliashberg, K. Gray, D. Gubser, D. Liebenberg, W. Little, K. Mueller, H. Piel, B. Raveau, D. Scalapino, I. Schuller, R. Schrieffer, R. Soulen and W. Pickett.

We are especially grateful to Vitaly Kresin and Mark Reeves for their invaluable help and many discussions.

The quality of manuscript has greatly benefited from the editorial skills of Jeffrey Robbins and Anita Lekhwani.

This book as well as our other scientific endeavors would not have been possible without the forebearance of Lilia Kresin, Iris Wolf and Terry Morawitz.

CONTENTS

Mechanisms of Conventional and High T_c Superconductivity

1

INTRODUCTION

1.1. Major goals

This book has several goals. There has been interest in high temperature superconductivity since well before the breakthrough in 1986 by Georg Bednorz and Alex Mueller [1] that started the recent revolution. In fact, from the theoretical standpoint there are many ways to achieve super-conductivity at very high temperatures, even well beyond what has been achieved experimentally.

Our first goal in writing this book was to present some of the theoretical routes to high temperature superconductivity independent of their experi-mental verification (see Chapters 2, 4, and 5 on phonon, electronic, and magnetic mechanisms). However, this book is *not* to be considered a detailed review of the field, it rather represents our own perspective. There have been many thousands of articles, and many review series and proceedings which can be used to gain a more complete picture of the realm of theories and experiments on high temperature superconductivity. A list of some of the review series and proceedings is provided as reference 2.

Some of the theoretical developments were motivated by the discoveries of the high-transition-temperature cuprates and are attempts to provide a mechanism for these materials. Of course, we have included a description of superconductivity mediated by the electron–phonon interaction (Chapter 2) and the experimental methods which were used to clearly establish this mechanism (Chapter 3). The phononic mechanism was the basis for the Bardeen–Cooper–Schrieffer (BCS) theory [3] which in its most general form [4] could account for almost all of the pre-1986 superconducting materials and can certainly account for high temperature superconductivity as well (see Chapter 2).

Our second goal was to present our theoretical viewpoint with regard to understanding the properties of the cuprate superconductors. This includes our view of the normal state properties, the superconducting properties independent of the underlying mechanism and finally the mech-anism itself. In addition we have tried to describe the current experimental state of the art, particularly those experiments that are most clearly associated with the mechanism (see Chapters 6 and 7).

1.2. Historical perspective

The story of superconductivity started in 1911 with the observation in Kamerlingh Onnes' laboratory that the resistance of mercury vanished below 5 K [5]. Of course that was the start of the race to discover new and higher transition temperature superconductors as well as to develop a theory to explain their properties. It took 50 years to develop a microscopic theory [3]; during this time the problem was worked on by many of the most prominent physicists of the twentieth century. There were many important discoveries during these years, among them the Meissner effect [6], whereby magnetic flux is excluded from the inside of a superconducting body, and the London equations [7], describing this phenomenon. During these years the maximum transition temperature had risen slowly (about 3 K per decade), reaching about 18 K in the late 1950s in both the B1 structure NbCN and the A15 structure Nb_3Sn.

An important theoretical advance came in 1950, with the development of the Ginzburg–Landau theory [8], which provided a phenomenological description of many of the thermodynamic, electromagnetic, and transport properties of the superconducting state. It was the solutions to these equations that provided the description of the magnetic vortices and the two fundamentally different types of superconductors distinguished by their very different behavior in a magnetic field. These two kinds of superconductors are Type I, which exhibits only the Meissner state, and Type II [9], which exhibits a mixed state where flux is not fully excluded but is incorporated into the superconductor in quantized flux entities called vortices.

Also in 1950, a major discovery, the isotopic shift in T_c, was made [10]. This discovery was one of the important clues to the microscopic mechanism since it showed that phonons were strongly involved in the superconductivity.

In 1957, a microscopic description of the superconducting state based on the phonon-mediated pairing of opposite-momentum, opposite-spin electrons on the Fermi surface was developed and published in the landmark paper by Bardeen, Cooper, and Schrieffer (BCS) [3]. This paper made many predictions about the behavior of this unique paired electron state that were almost immediately verified and provided the rapid acceptance of this theory. Even though this theory was approximate and was appropriate only for weak coupling of the electrons to the lattice, it worked remarkably well. The generalization to include the real phonon spectrum and strong electron–phonon coupling was provided by Eliashberg [4] (see Chapter 2).

In the early 1960s, superconducting junctions were shown to exhibit some additional very unusual properties. Both single-electron excitations (quasiparticles) and pairs could tunnel across a thin insulating barrier separating two superconducting electrodes. Quasiparticle tunneling was discovered and understood by Giaever [11a], whereas Josephson [11b] predicted the tunneling of pairs and the very unusual behavior of such junctions. In addition, the junctions provided the ultimate test of the generalized strong-coupling theory as well as providing a basis for determining the underlying mechanism (see Chapter 3).

In 1964, W. Little [12] described a model of high T_c organic polymers. This paper marked the beginning of the search for high T_c superconductivity and introduced the nonphonon, electronic mechanism as a key ingredient for such a search. This paper also predicted the phenomenon of organic superconductivity.

During this time many new superconducting materials were discovered and the list now totals more than 6000. Noteworthy among the pre-cuprate discoveries were discovery of the oxide superconductor LiTiO ($T_c = 13$ K) in 1973 [13], the oxide superconductor PbBaBiO in 1975 [14], the polymeric superconductor, SN_x ($T_c < 1$ K) in 1975 [15], and the organic superconductor TMTSF-PF$_6$ ($T_c = 1.5$ K) in 1980 [16]. Organic superconductors have now been synthesized with transition temperatures above 10 K.

The cuprates themselves have been very rich in new superconductors and there are presently more than 50 such compounds. The breakthrough came in 1986—75 years after the initial discovery of superconductivity! [1]. The La(Ba,Sr)CuO compounds studied by Raveau's group [17] were found to be superconducting by Bednorz and Mueller and have a minimum T_c of 40 K. Early in 1987, Wu, Chu and their colleagues reported on a YBaCuO compound with a transition temperature about 90 K, well above the temperature of liquid nitrogen [18]. This was followed very closely by reports of even higher transitions temperatures (above 100 K) in a BiSrCaCuO compound by Maeda et al. [19]. The highest transition temperature cuprate is a TlBaCaCuO compound with a transition temperature of above 125 K discovered by Herman and Sheng [20]. Of course the search continues and soon even higher transition temperatures may be found.

Very recently, superconductivity was discovered in a totally new class of compounds called the fullerenes [21]. These unusual carbon-based compounds are based on a soccerball-shaped molecule consisting of 60 carbon atoms. The molecules themselves form compounds with the alkali metals are then superconducting with transition temperatures approaching 30 K.

Thus superconductivity is a very rich field and we expect many new surprises in the future.

2

PHONON MECHANISM

2.1. Electron–phonon interaction

2.1.1. *The Hamiltonian*

In this section, we discuss the interaction between the conduction electrons and the crystal lattice in a metal. At first glance it may seem that it should be possible to treat this interaction in a manner analogous to that used in the theory of radiation, where the full Hamiltonian contains free electron and photon fields, plus a small interaction term. However, the present problem turns out to be quite delicate, the reason being that the very procedure of introducing the phonon field requires great care.

The starting point of the theory of metals is the adiabatic approximation introduced by Born and Oppenheimer [1]. This approximation is based on the fact that the masses of the ions and the conduction electrons are very different. The electrons adiabatically follow the motion of the ions. However, this adiabatic behavior is not perfect. The residual nonadiabaticity is a measure of the "friction" between the electrons and the lattice, representing what we refer to as the electron–phonon interaction.

The full Hamiltonian of the electron–ion system is

$$\hat{H} = \hat{T}_{\mathbf{r}} + \hat{T}_{\mathbf{R}} + V(\mathbf{r}, \mathbf{R}). \tag{2.1}$$

Here $\{\mathbf{r}\}$ and $\{\mathbf{R}\}$ are the sets of electronic and ionic coordinates, respectively, $\hat{T}$ is the kinetic energy operator, and $V(\mathbf{r}, \mathbf{R}) = V_1(\mathbf{r}) + V_2(\mathbf{R}) + V_3(\mathbf{r}, \mathbf{R})$ is the total potential energy (V_1 and V_2 correspond to the electron–electron and ion–ion interactions; V_3 describes the electron–ion interaction).

To proceed, we need to introduce the phonon subsystem. Furthermore, a rigorous description of the electron–phonon interaction should be based on the Hamiltonian being written in the form

$$\hat{H} = \hat{H}_0 + \hat{H}', \tag{2.2}$$

where

$$\hat{H}_0 = \hat{H}_0^{\mathrm{el}}(\mathbf{r}) + \hat{H}_0^{\mathrm{ph}}(\mathbf{R}) \tag{2.3}$$

and

$$\hat{H}' \equiv U(\mathbf{r}, \mathbf{R}) \tag{2.4}$$

Here $\hat{H}_0^{el}(\mathbf{r})$ and $\hat{H}_0^{ph}(\mathbf{R})$ describe the free electron and phonon fields, so that

$$\hat{H}_0^{el}(\mathbf{r}) = \hat{T}_r + U^{el}(\mathbf{r}), \tag{2.5}$$

$$\hat{H}_0^{ph}(\mathbf{R}) = \hat{T}_\mathbf{R} + U^{ph}(\mathbf{R}). \tag{2.6}$$

U^{el} and U^{ph} are the effective potential energies.

A rigorous theory has to be based on the representation (2.2), with the operator (2.4) corresponding to the electron–phonon interaction. The effects of this interaction can be calculated by perturbation theory (if H' is small), or with the help of the diagrammatic technique. However, it turns out that the problem of presenting the Hamiltonian (2.1) in the form (2.2), and, therefore, correctly introducing $\hat{H}'$ is quite nontrivial.

We will start out by looking at some conventional approaches to the problem of the electron–phonon interaction [1–3] (see also the reviews [4–7]). This will allow us to describe some general properties of adiabatic theory and, in addition, to stress the main difficulties with the introduction of an electron–phonon coupling Hamiltonian.

2.1.2. Adiabatic approximation. "Crude" approach

The total wave function of the system $\Psi(\mathbf{r}, \mathbf{R})$ is the solution of the Schrödinger equation

$$\hat{H}\Psi(\mathbf{r}, \mathbf{R}) = E\Psi(\mathbf{r}, \mathbf{R}). \tag{2.7}$$

The total Hamiltonian is defined by Eq. (2.1). Let us neglect, in the zeroth-order approximation, the term $\hat{T}_\mathbf{R}$ corresponding to the kinetic energy of heavy ionic system. Then we can introduce a complete orthogonal set of electronic wave functions which are the solutions of

$$[\hat{T}_r + V(\mathbf{r}, \mathbf{R})]\psi_m(\mathbf{r}, \mathbf{R}) = \varepsilon_m(\mathbf{R})\psi_m(\mathbf{r}, \mathbf{R}). \tag{2.8}$$

The functions $\psi_m(\mathbf{r}, \mathbf{R})$ and the eigenvalues $\varepsilon_m(\mathbf{R})$ (electronic terms) depend parametrically on the nuclear (ionic) positions.

The full wave functions $\Psi_n(\mathbf{r}, \mathbf{R})$ associated with the nth electronic state can be sought in the form of an expansion in the complete orthogonal set of the solutions of the electron Schrödinger equation (2.8). That is,

$$\Psi_n(\mathbf{r}, \mathbf{R}) = \sum_m \phi_{nm}(\mathbf{R})\psi_m(\mathbf{r}, \mathbf{R}). \tag{2.9}$$

Substituting the expansion (2.9) into Eq. (2.7), multiplying by $\psi_\alpha(\mathbf{r}, \mathbf{R})$, and integrating over **r** yields the following equation:

$$[\hat{T}_\mathbf{R} + \varepsilon_\alpha(\mathbf{R})]\phi_{n\alpha}(\mathbf{R}) = E\phi_{n\alpha}(\mathbf{R}). \tag{2.10}$$

We neglect the sum $\sum_m C_{\alpha m}\Phi_{nm}(\mathbf{R})$, where $C_{\alpha m}$ contains derivatives of ψ_n with respect to the nuclear coordinates.

According to Eq. (2.10), $\Phi_{n\alpha}(\mathbf{R})$ can be treated as a nuclear wave function; as for the electronic term, it appears as the potential energy for the motion of the heavy particles.

If we also neglect off-diagonal contributions to the expansion (2.9), we arrive at the Born–Oppenheimer expression for the total wave function:

$$\Psi_b(\mathbf{r}, \mathbf{R}) = \psi_n(\mathbf{r}, \mathbf{R})\phi_{nv}(\mathbf{R}), \tag{2.11}$$

where $\Phi_{nv}(\mathbf{R}) \equiv \Phi_{nnv}(\mathbf{R})$, v denotes the set of quantum numbers describing nuclear motion.

Equations (2.8), (2.10), and (2.11) define the Born–Oppenheimer adiabatic approximation. Inded, it is obvious from the derivation that the expression (2.11) is not an exact solution of the total Schrödinger equation. Let us evaluate the nonadiabaticity neglected in the course of deriving the approximation (2.11). The nonadiabaticity operator $\hat{H}'$ can be found in the following way (see, e.g., ref. 7). The zeroth-order operator $\hat{H}_0$, which has the function (2.11) as its eigenfunction, has the form

$$H_0 = \hat{T}_\mathbf{r} + V(\mathbf{r}, \mathbf{R}) + H_\mathbf{R}^0, \tag{2.12}$$

where the operator $\hat{H}_\mathbf{R}^0$ is defined by the relation

$$H_\mathbf{R}^0 \psi_n(\mathbf{r}, \mathbf{R})\phi_{nv}(\mathbf{R}) = \psi_n(\mathbf{r}, \mathbf{R})\hat{T}_\mathbf{R}\phi_{nv}(\mathbf{R}). \tag{2.13}$$

Indeed, one can see by direct calculation that

$$\begin{aligned}
H_0\Psi_{BO}(\mathbf{r}, \mathbf{R}) &= [\hat{T}_\mathbf{r} + V(\mathbf{r}, \mathbf{R})]\psi_n(\mathbf{r}, \mathbf{R})\phi_{nv}(\mathbf{R}) + H_\mathbf{R}^0\psi_n(\mathbf{r}, \mathbf{R})\phi_{nv}(\mathbf{R}) \\
&= \varepsilon_n(\mathbf{R})\psi_n(\mathbf{r}, \mathbf{R})\phi_{nv}(\mathbf{R}) + \psi_n(\mathbf{r}, \mathbf{R})\hat{T}_\mathbf{R}\phi_n(\mathbf{R}) \\
&= \psi_n(\mathbf{r}, \mathbf{R})[\hat{T}_\mathbf{R} + \varepsilon_n(\mathbf{R})]\phi_{nv}(\mathbf{R}) \\
&= E_{nv}\Psi_{BO}(\mathbf{r}, \mathbf{R}).
\end{aligned}$$

On the basis of Eqs. (2.2), (2.12), and (2.13), one can now determine the nonadiabaticity operator $\hat{H}'$:

$$H' = \hat{T}_\mathbf{R} - H_\mathbf{R}^0. \tag{2.14}$$

Hence

$$H'\Psi_{BO}(\mathbf{r}, \mathbf{R}) = \hat{T}_{\mathbf{R}}\psi_n(\mathbf{r}, \mathbf{R})\phi_{nv}(\mathbf{R}) - \psi_n(\mathbf{r}, \mathbf{R})\hat{T}_{\mathbf{R}}\phi_{nv}(\mathbf{R}), \qquad (2.15)$$

or, in matrix representation,

$$\langle mv'|H'|nv \rangle = \int \psi_m^*(\mathbf{r}, \mathbf{R})\phi_{mv'}^*(\mathbf{R})\hat{T}_{\mathbf{R}}\psi_n(\mathbf{r}, \mathbf{R})\phi_{nv}(\mathbf{R}) \, d\mathbf{R} \, d\mathbf{r}$$

$$- \delta_{mn} \int \phi_{mv'}^*(\mathbf{R})\hat{T}_{\mathbf{R}}\phi_{nv}(\mathbf{R}) \, d\mathbf{R}. \qquad (2.15a)$$

It is useful also to present the following classical estimate [8]. Equation (2.15) can be written in the form

$$H'\Psi_{BO}(\mathbf{r}, \mathbf{R}) = [H_{\mathbf{R}}, \psi_n]\phi_{nv}(\mathbf{R}). \qquad (2.15b)$$

If we assume that the motion of heavy particles is classical, then $i[\hat{H}_{\mathbf{R}}, \psi_n] \to (\partial\psi_n/\partial R)v$, where v is the velocity of the heavy particles. Thus the qualitative aspect of the adiabatic approximation is indeed connected with the small velocity of the ions relative to the much greater velocity of the electrons. Averaging the total potential energy $V(\mathbf{r}, \mathbf{R})$ over the fast electronic motion, we obtain the effective potential $\varepsilon(\mathbf{R})$ for the nuclear motion.

The adiabatic method is a rigorous approach which provides a foundation for the theory of solids. Nevertheless, from the practical point of view the approach described above has a definite shortcoming. Indeed, one can see from Eqs. (2.12) and (2.13) that the zeroth-order Hamiltonian $\hat{H}^0$ does not have the desired form (2.3).

Let us estimate the accuracy of the Born–Oppenheimer approximation. The characteristic scale of the nuclear wave function $\Phi_{nv}(\mathbf{R})$ is of the order of the vibrational amplitude a, whereas that of the electronic wave function is the lattice period L. As a result, we obtain

$$H'\Psi_{BO}(\mathbf{r}, \mathbf{R}) \sim -\frac{\hbar^2}{M}\frac{\partial\psi}{\partial R}\frac{\partial\phi}{\partial R}$$

$$\sim -\frac{\hbar^2}{M}\frac{\psi}{L}\frac{\phi}{a} \sim \frac{\hbar^2}{Ma^2}\frac{a}{L}\Psi_{BO} \sim \frac{a}{L}\hbar\omega\Psi_{BO},$$

that is,

$$H' \sim \frac{a}{L} \ll 1. \qquad (2.16)$$

Hence nonadiabaticity is proportional to the small parameter $\kappa = a/L$ (we have used the relation $a = [\hbar/M\omega]^{1/2}$).

Moreover, since $\omega = (K_n/M)^{1/2}$ where K is the force constant due to the Coulomb internuclear interaction, we can write $a^2 = \hbar/(K_n/M)^{1/2}$. Similarly, $L \cong \hbar/(K_e m)^{1/4}$. Since $K_e = K_n$ (the Coulomb interaction does not depend on the mass factor), we arrive at $\kappa \cong (m/M)^{1/4}$.

Another conventional form of adiabatic theory (the so-called "crude" or "clamped" approximation) is based on the fact that the ions are located near their equilibrium positions. In this version of the adiabatic theory, the expansion of the total wave function uses the electronic wave functions $\psi_n(\mathbf{r}, \mathbf{R}) \cong \psi_n(\mathbf{r}, \mathbf{R}_0)$, where $\mathbf{R}_0$ corresponds to some equilibrium configuration. That is,

$$\Psi_n(\mathbf{r}, \mathbf{R}) = \sum_m \tilde{\phi}_{nm}(\mathbf{R}) \psi_m(\mathbf{r}, \mathbf{R}_0). \qquad (2.17)$$

The total wave function in the zeroth-order approximation is $\Psi_n(\mathbf{r}, \mathbf{R}) = \psi_n(\mathbf{r}, \mathbf{R}_0) \tilde{\phi}_{nv}(\mathbf{R})$. Strictly speaking, the nuclear wave functions ϕ_{nv} and $\tilde{\phi}_{nv}$ are different (see Eq. (2.10)). The nonadiabaticity operator in the "crude" approximation has a relatively simple form:

$$H_1 = \sum_{1,\alpha} \frac{\partial V}{\partial X_{i\alpha|0}} \Delta X_{i\alpha}. \qquad (2.17a)$$

The "crude" approach has the same weakness as the Born–Oppenheimer approximation: the zeroth-order Hamiltonian does not have the form (2.3). As a result, neither approach is adaptable to the usual perturbation theory treatment. It is difficult to evaluate higher-order corrections or to consider the case of strong nonadiabaticity.

Now we turn to a different approach to adiabatic theory that will allow us to overcome these problems.

2.1.3. Electron–phonon coupling

We follow the approach developed by Geilikman [9, 10]. We introduce the functions

$$\Psi^0(\mathbf{r}, \mathbf{R}) = \psi_n(\mathbf{r}, \mathbf{R}_0) \phi_{nv}(\mathbf{R}), \qquad (2.18)$$

which form a complete basis set. The functions $\psi_n(\mathbf{r}, \mathbf{R})$ and $\phi_{nv}(\mathbf{R})$ are defined by the equation

$$[\hat{T}_\mathbf{r} + V(\mathbf{r}, \mathbf{R})] \psi_m(\mathbf{r}, \mathbf{R}) = \varepsilon_m(\mathbf{R}) \psi_m(\mathbf{r}, \mathbf{R})_{|\mathbf{R} = \mathbf{R}_0} \qquad (2.19)$$

and by Eq. (2.10).

Here n is the set of quantum numbers for the electronic states, v is the vibrational quantum number, $\mathbf{R}_0$ is the equilibrium configuration. Strictly speaking, $\mathbf{R}_0$ depends on n, but for levels close to the ground state this dependence may be neglected.

Equation (2.10) together with the wave functions $\phi_{nv}(\mathbf{R})$ serve to define phonons in the adiabatic theory.

We can write down the Hamiltonian in the zeroth-order approximation (the functions (2.18) are its eigenfunctions, see below, Eq. (2.21a)). Namely,

$$\hat{H}_0 = \hat{H}_{0\mathbf{r}} + \hat{H}_{0\mathbf{R}}, \tag{2.20}$$

where

$$\hat{H}_{0\mathbf{r}} = \hat{T}_{\mathbf{r}} + V(\mathbf{r}, \mathbf{R}_0), \tag{2.20a}$$

$$\hat{H}_{0\mathbf{R}} = \hat{T}_{\mathbf{R}} + \hat{A}(\mathbf{R}). \tag{2.20b}$$

The operators $\hat{T}_{\mathbf{r}}$ and $\hat{T}_{\mathbf{R}}$ are defined according to (2.1). The operator $\hat{A}$ is defined by the following matrix form:

$$\hat{A}_{nv'}^{n'v'} = \delta_{nn'} \int \phi_{n'v'}(\mathbf{R})[\varepsilon_n(\mathbf{R}) - \varepsilon_n(\mathbf{R}_0)]\phi_{nv}(\mathbf{R}) \, d\mathbf{R}. \tag{2.21}$$

It is easy to show that the operator $\hat{H}_0$ is diagonal in the space of the functions (2.18). Indeed,

$$\hat{H}_{0;nv}^{n'v'} = \int \psi_{n'}^*(\mathbf{r}, \mathbf{R}_0)\phi_{n'v'}(\mathbf{R})[\hat{T}_{\mathbf{r}} + V(\mathbf{r}, \mathbf{R}_0) + \hat{T}_{\mathbf{R}} + A(\mathbf{R})]$$

$$\times \psi_n(\mathbf{r}, \mathbf{R}_0)\phi_{nv}(\mathbf{R}) \, d\mathbf{r} \, d\mathbf{R}$$

$$= \delta_{nn'} \int \phi_{n'v'}(\mathbf{R})[\hat{T}_{\mathbf{R}} + \varepsilon_n(\mathbf{R})]\phi_{nv}(\mathbf{R}) \, d\mathbf{R}$$

$$= \delta_{nn'}\delta_{vv'}E_n. \tag{2.21a}$$

Note the important fact that the operator $\hat{H}_0$ represents a sum of two terms, one of which depends on $\mathbf{r}$, and the other on $\mathbf{R}$. These terms correspond to the electron and the phonon fields, respectively.

The functions (2.18) form the basis set in the present method. It is interesting that the total function (2.18) is a combination of the Born–Oppenheimer and the "crude" functions. Indeed, the electron wave function is that of the "crude" approach, whereas $\Phi_{nv}(\mathbf{R})$ is a solution of the nuclear Schrödinger equation of the Born–Oppenheimer version.

The electron–phonon interaction is given by the following nonadiabatic term in the full Hamiltonian:

$$\hat{H}' = \hat{H} - \hat{H}_0, \tag{2.22}$$

where $\hat{H}$ and $\hat{H}_0$ are defined by Eqs. (2.1) and (2.20). As a result, we find

$$\hat{H}' = V(\mathbf{r}, \mathbf{R}) - V(\mathbf{r}, \mathbf{R}_0) - \hat{A}(\mathbf{R}), \tag{2.23}$$

or, in matrix representation,

$$H_{na;\,nv}^{n'v'} = [V(\mathbf{r}, \mathbf{R}) - V(\mathbf{r}, \mathbf{R}_0)]|_{nv}^{n'v'} - \hat{A}(\mathbf{R})|_{nv}^{n'v'}. \tag{2.23a}$$

The last term is defined by Eq. (2.21).

The nonadiabatic Hamiltonian can be written as an expansion about the equilibrium coordinates $\mathbf{R}_0$:

$$H_{na;\,nv}^{n'v'} = \frac{\partial V}{\partial \mathbf{R}_{|0}} \delta\mathbf{R}|_{nv}^{n'v'} - \frac{\partial \varepsilon_n}{\partial \mathbf{R}_{|0}} \delta\mathbf{R}|_{nv}^{n'v'} + \frac{1}{2} \frac{\partial^2 V}{\partial \mathbf{R}_i \, \partial \mathbf{R}_{\kappa|0}} \delta\mathbf{R}_i \, \delta\mathbf{R}_\kappa|_{nv}^{n'v'}$$

$$- \frac{1}{2} \frac{\partial^2 \varepsilon_n}{\partial \mathbf{R}_i \, \partial \mathbf{R}_{\kappa|0}} \delta\mathbf{R}_i \, \delta\mathbf{R}_\kappa|_{nv}^{n'v'} + \frac{1}{6} \frac{\partial^3 V}{\partial \mathbf{R}_i \, \partial \mathbf{R}_\kappa \, \partial \mathbf{R}_{l|0}} \delta\mathbf{R}_i \, \delta\mathbf{R}_\kappa \delta\mathbf{R}_l|_{nv}^{n'v'} + \cdots. \tag{2.24}$$

The terms $\partial \varepsilon_n(\mathbf{R})/\partial \mathbf{R}|_0$ vanish. The first term corresponds to the "crude" approximation (see Eq. (2.17a)).

Note that the result (2.24) represents a power-series expansion in terms of the parameter (2.16).

Indeed, in Eq. (2.24) the deviation from the equilibrium position is $\delta R \simeq a$, while the equilibrium distance is $R_0 \simeq L$. It is convenient to rewrite Eq. (2.24) in the following way:

$$H_{na}|_{nv}^{n'v'} = H^{(1)}|_{nv}^{n'v'} + H^{(2)}|_{nv}^{n'v'} + \cdots \tag{2.25}$$

where

$$H^{(1)}|_{nv}^{n'v'} = \frac{\partial V}{\partial \mathbf{R}_{|0}} \delta\mathbf{R}|_{nv}^{n'v'} - \frac{\partial \varepsilon_n}{\partial \mathbf{R}_{|0}} \delta\mathbf{R}|_{nv}^{nv'} \tag{2.25a}$$

$$H^{(2)}|_{nv}^{n'v'} = \left[\frac{1}{2} \frac{\partial^2 V}{\partial \mathbf{R}_i \, \partial \mathbf{R}_{\kappa|0}} \delta\mathbf{R}_i \, \delta\mathbf{R}_\kappa - \frac{1}{2} \frac{\partial^2 \varepsilon_n}{\partial \mathbf{R}_i \, \partial \mathbf{R}_{\kappa|0}} \delta\mathbf{R}_i \, \delta\mathbf{R}_\kappa \right]_{nv}^{n'v'}. \tag{2.25b}$$

Let us estimate the magnitude of the terms $\hat{H}_1, \hat{H}_2, \ldots$. Starting with $\hat{H}_2$, we have $\hat{H}_2 \cong \hbar\omega$. Note that each additional factor $\Delta X_{i\alpha}$ brings in the parameter $a/L \ll 1$. Hence

$$H_1 \sim \hbar\omega/\kappa, \tag{2.26}$$

$$\tilde{H}_2 \sim \hbar\omega, \tag{2.26a}$$

$$H_3 \sim \kappa\hbar\omega, \quad \text{etc.} \tag{2.26b}$$

The fact that the first nonadiabatic terms ((2.26), (2.26a)) do not contain the smallness (they are not proportional to the small parameter κ) is an extremely important feature of the adiabatic theory. This is the reason why one must be careful in analyzing the effects of the electron–phonon coupling.

For example, one finds that the electron–phonon interaction makes only a very small contribution to the total energy of the system (see [9, 10]). At the same time, some other effects, such as mass renormalization, turn out to be very significant. The phenomenon of superconductivity is another example of a gigantic nonadiabatic effect.

2.1.4. Superconductivity as a nonadiabatic phenomenon

Superconductivity arises from an attractive interaction between electrons. This attraction is caused by the electron–lattice interaction, and its strength must be sufficient to overcome the Coulomb repulsion forces. The latter are actually somewhat weakened by a peculiar logarithmic factor (see below, Eq. (2.55)). Still, the interelectron attraction mediated by phonon exchange must be sufficiently strong: its energy must on the order of ε_F.

One may wonder whether the electron–phonon interaction is capable of producing such a large effect. It turns out that the answer is yes, and the reason has to do with the presence of massive nonadiabaticity.

Consider electrons near the Fermi level performing transitions between energy states in a layer of thickness $\sim \Omega_D$, so that $\Delta\varepsilon < \Omega_D$. Transitions of this kind are nonadiabatic, since the change in the electronic energy is smaller than the characteristic phonon energy. As we shall see, these nonadiabatic transitions are responsible for the strong electron–electron attraction.

Let us estimate the strength of the phonon-mediated electron–electron interaction. The matrix elements of interest, $M_{p_1,p_2}^{p_3,p_4}$, where $\mathbf{p}_1, \mathbf{p}_2$ are the initial electron momenta and $\mathbf{p}_3, \mathbf{p}_4$ are the final momenta, can be written in the form

$$M_{\mathbf{p}_1,\mathbf{p}_2;\mathbf{p}_3,\mathbf{p}_4} = \frac{|H_1|_{\mathbf{p}_1,\mathbf{p}_2;\, N_{\mathbf{q}\lambda}=0}^{\mathbf{p}_3,\mathbf{p}_2;\, N_{\mathbf{q}\lambda}=1} |H_1|_{\mathbf{p}_3,\mathbf{p}_2;\, N_{\mathbf{q}\lambda}=1}^{\mathbf{p}_3,\mathbf{p}_4;\, N_{\mathbf{q}\lambda}=0}}{\varepsilon_{\mathbf{p}_1} - \varepsilon_{\mathbf{p}_3} - \Omega_{\mathbf{q}\lambda}}$$

$$+ \frac{|H_1|_{\mathbf{p}_1,\mathbf{p}_2;\, N_{-\mathbf{q}\lambda}=0}^{\mathbf{p}_1,\mathbf{p}_4;\, N_{-\mathbf{q}\lambda}=1} |H_1|_{\mathbf{p}_1,\mathbf{p}_4;\, N_{-\mathbf{q}\lambda}=1}^{\mathbf{p}_3,\mathbf{p}_4;\, N_{-\mathbf{q}\lambda}=0}}{\varepsilon_{\mathbf{p}_2} - \varepsilon_{\mathbf{p}_4} - \Omega_{\mathbf{q}\lambda}}. \qquad (2.27)$$

The following conservation rules hold:

$$\mathbf{p}_3 - \mathbf{p}_1 = \mathbf{p}_2 - \mathbf{p}_4 = \mathbf{q}; \qquad \varepsilon_{\mathbf{p}_1} + \varepsilon_{\mathbf{p}_2} = \varepsilon_{\mathbf{p}_3} + \varepsilon_{\mathbf{p}_4},$$

where $\mathbf{q}$ is the phonon momentum.

Note that the expression (2.27) contains terms nondiagonal in the operator $\hat{H}_1$ (2.23). These nondiagonal terms are finite. Since we are considering virtual phonon exchange processes, the conservation of energy for the elementary act of electron–phonon scattering does not need to hold. The main contribution to the interaction comes from the nonadiabatic region. Unlike in the case of small molecules, the existence of such a region is guaranteed by the continuous nature of the electron and phonon spectra.

Let us estimate the matrix element (2.27) in this region. Using Eq. (2.26), we obtain

$$M \cong \frac{(\hbar\omega/\kappa)^2}{\hbar\omega} = \frac{\hbar\omega}{\kappa^2} = \frac{\hbar\omega}{(\hbar\omega/\varepsilon_F)} = \varepsilon_F. \qquad (2.28)$$

Hence the phonon-exchange interaction is very strong ($\sim \varepsilon_F$). This is due to the large contribution of $\hat{H}_1$, Eq. (2.26), which comes from the nonadiabatic domain $\Delta\varepsilon_{el} < \hbar\omega_D$.

The qualitative arguments presented above explain how a strong attraction can arise and overcome the Coulomb repulsion. A detailed quantitative analysis must be based on the methods of many-body theory, effects of this magnitude cannot be rigorously treated by perturbation theory.

2.2. Eliashberg equations. Superconductors with strong coupling

The BCS model [11] is based on a Hamiltonian with a 4-fermion interaction. Phonons as such are not explicitly present in this model; they exist only to set the energy scale for the cut-off of the interaction. Such an approach is permissible only if the electron–lattice interaction is weak. The situation is analogous to that found in quantum electrodynamics, where the degrees of freedom of the electromagnetic field can be excluded from the Lagrangian to second-order accuracy. Therefore, the 4-fermion model can be used if the electron–phonon interaction is weak, i.e., if the coupling constant $\lambda \ll 1$. This corresponds to the condition $T_c \ll \omega_D$.

In many superconductors this condition is not satisfied and $\lambda \gtrsim 1$. For example, in lead $\lambda = 1.4$, in mercury $\lambda = 1.6$, and in the alloy $Pb_{0.65}Bi_{0.35}$ $\lambda = 2.1$ (see, e.g., [12, 13]). In cases like this it is necessary to go beyond the limit of weak coupling. The Eliashberg equations [14], formulated shortly after the creation of the BCS theory, allow to analyze the properties of such superconductors.

The Eliashberg equations are based on the many-body theory and are a generalization of the work of A. Migdal on the properties of normal metals [15]. A detailed description of the fundamentals of these methods can be found in a number of reviews and monographs (see, e.g., [5, 16–18]). Below, we outline the derivation of the Eliashberg equations on the basis of the Feynman diagram technique.

2.2.1. Self-energy parts

Let us start by considering a metal at $T = 0$ K. An electron moving through the crystal lattice polarizes the latter. Of course, this polarization also acts

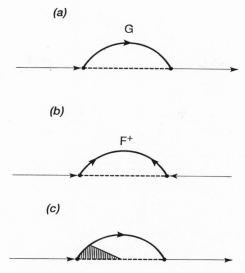

FIG. 2.1. Self-energy parts.

back on the electron and affects its motion. In quantum language, polarization can be described as emission and subsequent absorption of virtual phonons. The self-energy part Σ_1, depicted in Fig. 2.1a, describes this process. Here G is the electron Green's function,

$$D = \frac{\Omega(q)}{2} \{[\omega - \Omega(q) + i\delta]^{-1} - [\omega + \Omega(q) - i\delta]^{-1}\} \qquad (2.29)$$

and Ω is the phonon frequency.

The self-energy part represents a certain effective potential in which the electron finds itself moving.

The electron–phonon interaction can lead not only to scattering, described by Σ_1, but also to superconductive pairing. This pairing is due to the electron–electron attraction caused by phonon exchange. This attraction leads to an instability of the normal state [19]. The instability is manifested in the appearance of an imaginary pole in the two-particle Green's function (see, e.g., [16]).

Pairing is described by the self-energy part Σ_2, which has the form given in Fig. 2.1b. Here D is the phonon Green's function (2.29), and F^+ is the anomalous Green's function introduced by Gor'kov [20], which describes Cooper pairing. It is defined as follows:

$$F^+(x_1, x_2) = -i\langle N + 2|T\psi^+(x_1)\psi^+(x_2)|N\rangle;$$

T is the time-ordering operator, N is the number of particles.

FIG. 2.2. Diagrammatic equations for G and F^+ functions.

The Green's functions G and F^+ are given by the diagrammatic equations in Fig. 2.2. By analyzing these equations, one can obtain an integral equation for Σ_2 or, as turns out to be more convenient, for the function $\Delta(\mathbf{p}, \omega) = \Sigma_2 Z^{-1} = \Sigma_2 (1 - f/\omega)^{-1}$. Here $f(\omega)$ is the odd part of the self-energy function Σ_1.

Let us consider an isotropic model and set $p \cong p_F$. Then we can change the variables of integration to the energy ξ' which is referred to the Fermi level and phonon momenta q. Carrying out the integration over ξ' and going over to integration over the phonon frequencies, we arrive at the following system of equations:

$$\Delta(\omega) = [Z(\omega)]^{-1} \int_0^{\omega_c} d\omega' \, P(\omega')[K_+(\omega, \omega') - \mu^*], \tag{2.30}$$

$$[1 - Z(\omega)]\omega = \int_0^\infty d\omega' \, N(\omega')K_-(\omega, \omega'), \tag{2.31}$$

where

$$P(\omega) = \text{Re}\left\{\frac{\Delta(\omega)}{[\omega^2 - \Delta^2(\omega)]^{1/2}}\right\}, \tag{2.32}$$

$$K_\pm(\omega, \omega') = \int_0^\infty d\Omega \, \alpha^2(\Omega)F(\Omega)\left(\frac{1}{\omega' + \omega + \Omega + i\delta} \pm \frac{1}{\omega' - \omega + \Omega - i\delta}\right). \tag{2.33}$$

These equations contain an important function $g(\omega) = \alpha^2(\Omega)F(\Omega)$, where $F(\Omega)$ is the phonon density of states:

$$F(\Omega) = \sum_\lambda \int \frac{d\mathbf{p}}{(2\pi\hbar)^3} \, \delta(\Omega - \Omega_{\mathbf{p}\lambda}) \tag{2.34}$$

(we are summing over all polarizations λ), and $\alpha^2(\Omega)$ describes the electron–

phonon coupling. A rigorous definition of $g(\omega)$ reads as follows:

$$\alpha^2(\Omega)F(\Omega) = \frac{\int \frac{dS_{k'}}{|\boldsymbol{v}_{k'}|} \int \frac{dS_k}{|\boldsymbol{v}_k|} \frac{1}{(2\pi)^3\hbar} \sum_{\lambda} |g_{k',k,\lambda}|^2 \, \delta[\Omega - \omega_{\lambda,k'-k}]}{\int dS_k/|\boldsymbol{v}_k|}. \qquad (2.35)$$

Equations (2.30) and (2.31) are the Eliashberg equations.

2.2.2. General properties of the Eliashberg equations

The main Eliashberg equation is Eq. (2.30) which defines the function $\Delta(\omega)$, known as the order parameter. This function should not be confused with the energy gap. The latter is defined by the equation

$$N(\omega) = \text{Re}\left\{\frac{|\omega|}{[\omega^2 - \Delta^2(\omega)]^{1/2}}\right\}, \qquad (2.36)$$

where $N(\omega)$ is the density of states in the superconducting state; the density of states is normalized to unity in the normal state. The energy gap corresponds to the region where the electron density of states vanishes.

Equations (2.30–2.31) have a complicated mathematical structure. The equation for the order parameter $\Delta(\omega)$ is a nonlinear integral equation. For a given form of the phonon spectrum, i.e., for any particular phonon density of states function $F(\Omega)$ and a particular electron–phonon interaction, this equation yields the real and imaginary parts of $\Delta(\omega)$.

A very important feature of the Eliashberg equations is that the integrand contains only the product of two Green's functions and the matrix elements of the electron–phonon interaction. Generally speaking, one could expect that the total vertex would also be present (see Fig. 2.1c). However, to within the adiabatic accuracy ($\kappa \ll 1$, or $m/M \ll 1$), one can keep only the lowest order [15] (see also [16, 17]).

Equations (2.30) and (2.31) hold for $T = 0$ K. The case of finite temperatures, which is especially important since it has to do with the calculation of T_c, can be analyzed with the help of the method of thermodynamic Green's functions. This method, (see, e.g., [16, 18]) turns out to represent the most effective approach. In this formalism the Eliashberg equations have the following form:

$$\Delta(\omega_n)Z = \pi T \sum_{n'} \int d\Omega \frac{g(\Omega)}{\Omega} D(\Omega_n - \omega_{n'}, \Omega) \frac{\Delta(\omega_{n'})}{|\omega_{n'}|}, \qquad (2.37)$$

$$g(\Omega) = \alpha^2(\Omega)F(\Omega),$$

and a similar equation holds for Z. In Eq. (2.37),

$$D = \Omega^2[\Omega^2 + (\omega_n - \omega_{n'})^2]^{-1}. \tag{2.38}$$

We have written this equation for the case $T = T_c$. For $T < T_c$, the substitution $|\omega_n| \rightarrow [\omega_n^2 + \Delta^2(\omega_n)]^{1/2}$ has to be made in the denominator. In Eqs. (2.37), $\omega_n = (2n + 1)\pi T$.

The function $\Delta(\omega_n)$ which appears in (2.37) is called the thermodynamic order parameter. The usual order parameter $\Delta(\omega)$ which appears in (2.34) can be obtained by analytically continuing $\Delta(\omega_n)$ into the upper half-plane.

We introduce the quantity

$$\lambda = 2 \int \frac{d\Omega}{\Omega} g(\Omega), \tag{2.39}$$

which we call the coupling constant. Note that the coupling constant does not directly appear in Eqs. (2.30) and (2.37). It arises explicitly, for example, if we consider a model with a single phonon frequency, e.g., a delta function-like peak with $\alpha^2 F = (\lambda\Omega/2)\delta(\Omega - \Omega_1)$.

We can also consider the case of two peaks. In this case the right-hand side of the Eliashberg equation will consists of two terms each containing the coupling constant of a single peak:

$$\alpha^2(\Omega)F(\Omega) = \frac{\lambda_1\Omega_1}{2}\delta(\Omega - \Omega_1) + \frac{\lambda_2\Omega_2}{2}\delta(\Omega - \Omega_2). \tag{2.40}$$

In this case $\lambda = \lambda_1 + \lambda_2$.

In real life one deals with complicated phonon spectra and the superconducting parameters correlate directly with the function $\alpha^2(\Omega)F(\Omega)$. Strictly speaking, it is not possible to rewrite Eqs. (2.30) and (2.37) in a form which would explicitly contain the coupling constant.

Nevertheless, the coupling constant concept remains very useful. This is due to the differences in the behavior and the temperature dependences of the phonon Green's function (2.38) and of the function $g(\Omega) = \alpha^2(\Omega)F(\Omega)$. The Green's function D is a relatively smooth function of frequency. The case is usually quite different for the function $g(\Omega)$ and especially for the phonon density of states factor $F(\Omega)$. The latter typically contains a number of rather sharp peaks whose origin is as follows.

The phonon spectrum of a solid consists of several branches corresponding to longitudinal and transverse acoustic, and optical phonons. At small wave vectors q, the acoustic branches have simple linear dispersion laws, e.g., $\Omega_{tr} = u_{tr}q$. For large q, on the other hand, there are deviations from the linear behavior giving rise to high-density-of-states regions (van Hove singularities). In these regions the frequency varies only weakly with the momentum with the result that the density of states becomes very high, as we have stated. These regions, where the function $g(\Omega) = \alpha^2(\Omega)F(\Omega)$ contains peaks, make the strongest contribution to electron pairing.

Let us now consider the factor $g(\Omega)D(\Omega)/\Omega$. Breaking up the integration over frequencies into separate intervals containing different peaks, and making use of the smoothness of the phonon Green's function, we can write the right-hand side of the Eliashberg equation as a sum of terms of the form $\lambda_i D(\Omega_i)$, where the coupling constants are defined by expressions of the type (2.39), each referring to a different peak.

For many materials one can introduce an effective average phonon frequency, thereby casting the Eliashberg equation into a form corresponding to a single characteristic frequency and a single coupling constant defined by (2.39). According to [21], the function

$$K_{n-n'} = 2 \int d\Omega \, g(\Omega)\Omega \times [\Omega^2 + (n - n')^2(2\pi T)^2]^{-1}$$

can be replaced by

$$K_{n-n'} \equiv K^0_{n-n'} \approx \lambda v^2[v^2 + (n - n')^2]^{-1},$$

where

$$v = \tilde{\Omega}/2\pi T, \qquad \tilde{\Omega} = \langle \Omega^2 \rangle^{1/2},$$

$$\langle \Omega^n \rangle = (2/\lambda) \int g(\Omega)\Omega^{n-1} \, d\Omega$$

with high accuracy: $K_n = K^0_n(1 - r)$ with $r \ll 1$ ($r \propto \delta = [\tilde{\Omega}/\langle \Omega \rangle] - 1$).

The limit of weak coupling deserves special attention, both for its own sake and in order to establish contact with the usual formalism of the BCS theory. This limit corresponds to $\lambda \ll 1$. As can be shown self-consistently from the results of weak-coupling theory, this means that $\pi T_c \ll \tilde{\Omega}$ where $\tilde{\Omega}$ corresponds to the important short-wavelength part of the phonon spectrum. In the weak-coupling approximation we may set $Z \cong 1$. Furthermore, we can neglect the dependence of the phonon Green's function on ω_n [see below; this approximation is accurate to $\sim (T_c/\tilde{\Omega})^2$]. Then $\Delta(\omega_n)$ also becomes independent of ω_n, and we arrive at the equation

$$1 = \lambda \sum_n D(\omega_n, \tilde{\Omega})/|\omega_n|. \tag{2.41}$$

The function D guarantees the convergence of the summation at the upper limit by effectively cutting off this summation at $\omega_{n'} \sim \tilde{\Omega}$. To calculate T_c we can either directly employ Eq. (2.41) (see Section 2.3.1), or we can approximate $D \cong 1$ and terminate the summation at $\tilde{\Omega}$. This results in the well-known equation of the BCS theory: $T_c(\Omega) \cong \tilde{\Omega} \exp(-1/\lambda)$.

The weak-coupling theory is preexponentially accurate: the approximation $D \cong 1$ affects the preexponential factor. Hence the introduction of the coupling constant λ is valid to the same accuracy. We will discuss the weak coupling case in more detail in the next section.

2.3. Critical temperature

The Eliashberg equation is valid for any strength of the electron–phonon interaction. On the basis this equation, let us begin by evaluating the principal parameter of the theory: the critical temperature.

The value of T_c is determined, by several quantities: $T_c = T_c(\lambda, \mu, \tilde{\Omega})$, where λ describes the strength of the electron–phonon coupling, $\tilde{\Omega}$ stands for the characteristic phonon frequencies and sets the energy scale (in a rough approximation, $\tilde{\Omega} \simeq \Omega_D$, where Ω_D is the Debye temperature), and μ^* describes the Coulomb repulsion. Curiously, it will turn out that the specific forms of the analytic expressions for T_c are different for different coupling strengths.

2.3.1. Weak coupling

We consider first the case of weak electron–phonon interaction. This case corresponds to the original BCS model.

In this section we employ the Eliashberg equation (2.37) to consider the weak coupling base. This approach allows us to avoid forcing a cut-off at the characteristic phonon energy (it will take place automatically); in addition, the preexponential accuracy of the model becomes immediately apparent.

Equation (2.37) serves as our starting point [22]. The expression for the constant g^2 describing the electron–phonon interaction can be written in the following form [23]:

$$g^2 = \zeta \frac{u^2 q^2}{\Omega_j^2(q)} \gamma_j(\mathbf{q}). \tag{2.42}$$

Here u is the speed of sound, ζ is the Fröhlich parameter which is directly related to the electron–phonon matrix element [3, 10], q is the phonon momentum, and $\gamma(q) \simeq 1$. It is useful to separate out the dependence of the electron–phonon matrix element on the phonon momentum and frequency, see, e.g., [6]. Note the important fact that g depends on the phonon frequency; this dependence vanishes only for a purely acoustic dispersion law. For simplicity, we are not explicitly including the Coulomb interaction; it can be taken into account in the usual manner (see below). In Eq. (2.42) we are summing over all phonon branches j. Henceforth we omit the index j.

From Eq. (2.42), after going over to integration over the phonon momentum and ξ, we obtain

$$\Delta(\omega_n, T) = \frac{\zeta T}{2p_0^2} \int_0^{k_1} q \, dq \, \frac{u^2 q^2}{\Omega^2(q)} \gamma(q) \sum_{n'} \int d\xi \, \frac{\Omega^2(q)}{\Omega^2(q) + (\omega_n - \omega_{n'})^2}$$

$$\times \frac{\Delta(\omega_{n'}, T)}{\omega_{n'}^2 + \xi^2 + \Delta^2(\omega_{n'}, T)}, \tag{2.43}$$

$$k_1 = \min\{2p_F, q_{\max}\}.$$

In the weak-coupling approximation $\zeta \ll 1$, and as a result one can neglect the direct dependence of $\Delta(\omega_n, T_c)$ on ω_n and take $\Delta(\omega_n, T_c) \simeq \varepsilon(T)$, where $\varepsilon(T)$ is the energy gap (the correction turns out to be of the order of $\lambda T_c^2/\Omega_D^2$; see [22]). Since $\varepsilon(T = T_c) = 0$, the critical temperature is determined by the equation

$$1 = \frac{\zeta T}{2p_0^2} \int_0^{k_1} q \, dq \, \frac{u^2 q^2}{\Omega^2(q)} \gamma(q) \sum_{n'} \int d\xi \, \frac{\Omega^2(q)}{\Omega^2(q) + \omega_{n'}^2} \, \frac{1}{\omega_{n'}^2 + \xi_{|T=T_c}^2}. \tag{2.44}$$

The BCS model is recovered if in Eq. (2.44), in addition to assuming acoustic dispersion, we replace the phonon Green's function by unity in the interval $0 < \omega_n < \Omega_D$. In the approach based on the Eliashberg equation, the presence of the D-function automatically removes the logarithmic divergence.

Now summing over ω_n and integrating over ξ, we find

$$1 = \frac{\zeta}{2p_0^2} \int_0^{k_1} q \, dq \, \frac{u^2 q^2}{\Omega^2(q)} \gamma(q) \ln \frac{2\Omega(q)\gamma}{\pi T_k}. \tag{2.44a}$$

Introducing an auxiliary constant $\tilde{\Omega}$ such that

$$\ln \frac{2\Omega(q)\gamma}{\pi T_k} = \ln \frac{2\Omega(q)\gamma}{\pi \tilde{\Omega}} - \ln \frac{\tilde{\Omega}}{T_k} \tag{2.44b}$$

we arrive, with the help of (2.44a), at the following expression for the critical temperature:

$$T_k = \tilde{\Omega} \exp\left(-\frac{a+1}{\lambda}\right), \tag{2.45}$$

$$a = \frac{\zeta}{2p_0^2} \int_0^{k_1} q \, dq \, \frac{u^2 q^2}{\Omega^2(q)} \gamma(q) \ln \frac{2\gamma\Omega(q)}{\pi \tilde{\Omega}}, \tag{2.45a}$$

$$\lambda = \frac{\zeta}{2p_0^2} \int_0^{k_1} q \, dq \, \frac{u^2 q^2}{\Omega^2(q)} \gamma(q). \tag{2.45b}$$

Note that T_c does not depend on the choice of $\tilde{\Omega}$, which can therefore be picked arbitrarily. We choose $\tilde{\Omega}$ so as to minimize a (for instance, for acoustic dispersion $\tilde{\Omega} \simeq \Omega_D$, while in the case of a sharp peak at $\Omega = \Omega_1$ in the phonon density of states, $\tilde{\Omega} = 2\gamma\Omega_1/\pi$). This leads us to the following result:

$$T_c = \tilde{\Omega} \exp(-1/\lambda), \tag{2.46}$$

where λ is given by Eq. (2.45b).

Equation (2.46) is the result for T_c in the weak-coupling limit ($\lambda \ll 1$). It is clear that this result is preexponentially accurate. Indeed, the presence of the factor a in Eq. (2.45) is equivalent to a change in the preexponential factor.

Generally speaking, the critical temperature is very sensitive to the character of the phonon spectrum; the dominant contribution comes from the frequency dependence of the denominator in the exponent.

Note that it is only if the dispersion is purely acoustic, $\Omega = uq$ (in this case $\gamma = 1$), that the exponent does not depend on the phonon frequency. Indeed, in this case (with $2p_F < q_D$) Eq. (2.45b) gives

$$T_c = \tilde{\Omega} \exp(-1/\zeta) \tag{2.47}$$

(we have used the fact that $\zeta = \beta p_F$). Thus the usual BCS expression is recovered only in the weak-coupling approximation and, furthermore, only under the assumption of an acoustic dispersion law.

On the other hand, if the dispersion law $\omega(q)$ is not acoustic over the entire range of q, as is the case in real materials, then according to (2.45b) and (2.46) T_c is very sensitive to the form of the function $\Omega(q)$. Here this sensitivity is more pronounced than in the case of strong coupling (see below), because in the weak-coupling regime renormalization effects plays only a small role.

If the phonon density of states is described by a delta function-like peak at a frequency Ω_1, the effective coupling constant is $\lambda \sim \Omega_1^{-2}$.

2.3.2. *Intermediate coupling* ($\lambda \lesssim 1.5$)

In the preceding section we discussed the case of weak coupling ($\lambda \ll 1$). If the superconductor is characterized by a larger coupling constant ($\lambda \cong 1$), the approximations employed there are no longer valid. In particular, we can no longer neglect the dependence of Δ on ω_n. An analysis of this case was carried out by McMillan [24]. McMillan made use of the Eliashberg equation and computed the critical temperature for a phonon spectrum of the type found in niobium. This phonon spectrum [or, more precisely, the phonon density of states $F(\Omega)$] is known from neutron spectroscopy data, and has a structure typical of many metals. In addition, McMillan neglected the dependence of α on Ω and assumed that T_c can be expressed as a function of λ. This numerically calculated T_c can be analytically fitted by the formula

$$T_c = \frac{\Theta_D}{1.45} \exp\left(-\frac{1.04}{\rho - \dfrac{\mu^*}{1+\lambda} - 0.62\mu^*\rho}\right). \tag{2.48}$$

Here μ^* is the Coulomb pseudopotential (see Section 2.2.3), $\rho = \lambda(1+\lambda)^{-1}$, and $\Theta = \Omega_D$.

Dynes [25] modified the McMillan equation, replacing the Debye temperature by a characteristic frequency $\langle\Omega\rangle$; the average is defined by

$$\langle f(\Omega)\rangle \equiv \frac{2}{\lambda}\int_0^\infty \alpha^2 F(\Omega) f(\Omega)\Omega^{-1}\, d\Omega \tag{2.49}$$

The equation for T_c then reads

$$T_c = \frac{\langle\Omega\rangle}{1.2}\exp\left(-\frac{1.04(1+\lambda)}{\lambda - \mu^*(1+0.62\lambda)}\right). \tag{2.50}$$

We have already mentioned that the McMillan equation was obtained as a fit to the numerical solution for a niobium-like phonon spectrum; this equation has been applied with great success to many superconductors (see, e.g., [12, 13]). An analytical calculation for the model of a single delta function-like peak was carried out in [26] and has led to the expression which is close to the McMillan–Dynes formula (2.50) [see below, Eq. (2.54)]. This means that Eq. (2.50) is valid if $\pi T_c < \tilde{\Omega}$, that is, if $\lambda \lesssim 1.5$.

McMillan also obtained an interesting expression for the electron–phonon coupling constant:

$$\lambda = \frac{N(0)\langle I^2\rangle}{M\langle\Omega^2\rangle}, \tag{2.51}$$

[see Eq. (2.2)], where $\langle I\rangle$ is the average matrix element of the electron–phonon interaction, $N(0)$ is the density of electronic states, and M is the ion mass. It is important that the dependence of T_c on the phonon frequency is not determined solely by the preexponential factor: in principle, a strong dependence is also contained in the exponent itself. Note that these two dependences have opposite effects on T_c. This fact has led McMillan to propose a softening mechanism for increasing T_c. The previously accepted notion, based on the original BCS theory, was that a rise in the phonon frequency serves only to increase the preexponential factor and, consequently, increases T_c. The dependence (2.51) shows that a rise in the phonon frequency may, generally speaking, lead to the opposite result as well.

As was mentioned above, the McMillan equation is valid for $\lambda \lesssim 1.5$. In connection with this fact, it is interesting to note that from the expression

$\lambda \cong a\langle\Omega\rangle^{-2}$ [here $a = \langle I\rangle N(0)vM^{-1}$, see Eq. (2.51)] and the formula (2.50) it appears that there exists an upper limit on the attainable value of T_c; this was pointed out in the original McMillan paper. Indeed, neglecting μ^* for simplicity and calculating $\partial T_c/\partial\langle\Omega\rangle$ we find $T_c^{max} \cong \Omega_m \exp(-3/2)$, where $\Omega_m \cong (a/2)^{1/2}$]. This maximum value corresponds to $\lambda \cong 2$.

This conclusion, however, has to be taken with a great deal of caution. The fact of the matter is that even the value $\lambda = 2$ is outside the range where the McMillan equation is applicable; this is even more so for higher values of λ. We will come back to this point later when we are discussing the expression for T_c which holds for large values of λ.

2.3.3. Coulomb interaction

In addition to the attraction mediated by phonon exchange, we must keep in mind the presence of the Coulomb repulsion. It turns out that an important aspect of this problem is the phenomenon of logarithmic weakening of the repulsion. This factor has to do with the difference in the energy scales of the attraction and repulsion effects. The attraction is important in an energy interval $\sim\tilde{\Omega} \cong \Omega_D$. The repulsion, on the other hand, is characterized by the electronic energy scale $\sim\varepsilon_F$.

Let us look at this problem in more detail [24, 26–30]. The equation for the order parameter $\Delta(\omega_n)$ in the presence of Coulomb repulsion can be written as

$$\Delta(\omega_n) = \frac{1}{1 - f^n/\omega_n} \pi T \sum_{\omega_{n'}} [\lambda D(\omega_n - \omega_{n'}, \tilde{\Omega}) - \tfrac{1}{2}V_c\Theta(\varepsilon_0 - |\omega_{n'}|)] \frac{\Delta(\omega_{n'})}{|\omega_{n'}|}$$

$$(2.52)$$

The second term in square brackets is the one describing the Coulomb force ($\varepsilon_0 \cong \varepsilon_F$). Recall that the Green's function D provides a cut-off at energies $\sim\tilde{\Omega}$. We can seek the solution in the following form:

$$\Delta(\omega_n) = \Delta_0 \frac{\tilde{\Omega}^2}{\tilde{\Omega}^2 + \omega_n^2} + \Delta_\infty \frac{\omega_n^2}{\tilde{\Omega}^2 + \omega_n^2}. \qquad (2.53)$$

The second term reflects the presence of the repulsion. It results in the order parameter continuing outside the region $\sim\tilde{\Omega}$.

The quantities Δ_0 and Δ_∞ are defined by

$$\Delta_0 = \rho\pi T \sum_{\omega_{n'}} \frac{\tilde{\Omega}^2}{\tilde{\Omega}^2 + \omega_{n'}^2} \frac{\Delta(\omega_{n'})}{|\omega_{n'}|} - \frac{V_c}{1 + \lambda} \pi T \sum_{\omega_{n'}}^{\varepsilon_0} \frac{\Delta(\omega_{n'})}{|\omega_{n'}|},$$

$$(2.53a)$$

$$\Delta_\infty = -V_c \pi T \sum_{\omega_{n'}} \frac{\Delta(\omega_{n'})}{|\omega_{n'}|}.$$

Substituting Eq. (2.53) and eliminating the constants Δ_0 and Δ_∞, we arrive at the following result for T_c:

$$T_c = 1.14\tilde{\Omega} \exp\left(-\frac{1 + 0.5\rho - 0.35\rho^2 + 0.8\rho[\mu^*/(1 + \lambda)] + 0.4\mu^*\rho^2}{\rho - (\mu^*/1 + \lambda)] - 0.5\rho\mu^* - 1.5\mu^*\rho^2}\right). \quad (2.54)$$

Here

$$\mu^* = \frac{V_c}{1 + V_c \ln \varepsilon_0/\tilde{\Omega}} \quad (2.55)$$

is the so-called Coulomb pseudopotential. It contains the large logarithmic factor $\sim \ln(\varepsilon_0/\Omega)$ which reduces the contribution of the Coulomb repulsion. Usually $\mu \cong 0.1$–0.15, although its value might be different. For example, $\mu = 0.2$ for Nb_3Sn. On the other hand, $\mu^* \simeq 0$ for YB_6 [31].

Note that if the superconductor contains, in addition to the phonon subsystem, some high-energy electronic excitations which provide an additional attraction, this leads to an effective decrease in μ^*. We will discuss this question in more detail in Chapters 3 and 6.

2.3.4. Very strong coupling

We now consider the case of very strong coupling ($\lambda \gg 1$). We shall see that this regime corresponds to the criterion $2\pi T_c \gg \Omega$. This case was first analyzed in [12] by means of numerical calculations (for $\mu^* = 0$). An analytical treatment, based on a direct solution of the Eliashberg equation, was presented in [32].

It is convenient to use the dimensionless form of the Eliashberg equation. It turns out that for $\lambda \gg 1$ the function $\Delta(n)$ decays rapidly. In this case the matrix method developed by Owen and Scalapino [33] is extremely useful. The general equation which determines T_c can be written in the form

$$\phi_n = \sum_{m \geq 0} \tilde{K}_{nm} \phi_m, \quad (2.56)$$

$$\tilde{K}_{nm} = \frac{\alpha}{(2n + 1)^{1/2}(2m + 1)^{1/2}} F_{n,m,v}, \quad (2.56a)$$

where

$$F_{n,m,v} = \frac{1}{v^2 + (n - m)^2} + \frac{1}{v^2 + (n + m + 1)^2} - \delta_{nm} \sum_{l=0}^{2n} \frac{1}{v^2 + (n - l)^2}, \quad (2.56b)$$

where $\phi_n = \Delta_n/(2n + 1)^{1/2}$, $v = \tilde{\Omega}/2\pi T_c$, $\tilde{\Omega} = \langle \Omega^2 \rangle^{1/2}$, $\alpha = \lambda v^2$, and λ is defined, as usual, by Eq. (2.40). Equation (2.56) can be obtained directly from Eq. (2.37) and the equation for Z.

It should be pointed out that the presence of the renormalization function Z [corresponding to the last term on the right-hand side of Eq. (2.56)] plays an important role: it results in a cancellation of the diagonal term $n = m$.

Equation (2.56) is valid for any λ. The case of very strong coupling corresponds to $v \ll 1$, and therefore the quantity v can be neglected.

In the zeroth-order approximation, only $\phi_0 \neq 0$. Then Eq. (2.56) gives $\tilde{K}_{00} = 1$, where, according to Eqs. (2.56a) and (2.56b) $K_{00} = \alpha$. We then obtain

$$T_{c0} = (2\pi)^{-1}\lambda^{1/2}\tilde{\Omega}, \qquad \tilde{\Omega} = \langle \Omega^2 \rangle^{1/2}. \tag{2.57}$$

In the next approximation we keep more terms. Specifically, we solve the equation $\det \tilde{M} = 0$, where the matrix $\tilde{M}$ is defined by $\tilde{M} = \tilde{K} - \tilde{1}$. Then $\tilde{M}_{00} = \alpha - 1$, $\tilde{M}_{01} = \tilde{M}_{10} = 0.72\alpha$, and $\tilde{M}_{11} = -0.63\alpha - 1$. After a simple calculation, we obtain $\alpha = 0.785$, and based on Eq. (2.56) we finally arrive at the expression

$$T_c = 0.18\lambda^{1/2}\tilde{\Omega}. \tag{2.58}$$

The next iteration changes the coefficient in Eq. (2.58) by only a negligible amount. It is remarkable that even the zeroth-order approximation describes T_c with good accuracy ($\simeq 13\%$). Equation (2.58) is valid for $\lambda \gtrsim 5$.

It is interesting to consider the effect of the Coulomb interaction on T_c in the limit of strong coupling. The Eliashberg equation can be reduced to the form [cf. Eq. (2.56)]

$$\phi_n = \sum_m \tilde{K}^c_{nm} \phi_m,$$

where

$$\tilde{K}^c_{nm} = (2n + 1)^{-1/2}(2m + 1)^{-1/2}F^c_{n,m,v}|_{v=0}. \tag{2.59}$$

Solving Eq. (2.59) by analogy with the case $\mu^* = 0$, we obtain

$$T_c = 0.18\lambda_{eff}^{1/2}\tilde{\Omega}, \tag{2.60}$$

$$\lambda_{eff} \simeq \lambda(1 + 2.6\mu^*)^{-1}. \tag{2.60a}$$

Since $\mu \ll 1$, we can restrict ourselves to the linear term.

We see from Eq. (2.60) that the Coulomb term decreases the effective constant ($\lambda_{eff} < \lambda$), but this decrease differs in a striking way from that in the weak-coupling approximation. In the latter case, the effective constant has the well-known form $\lambda_{eff} = \lambda - \mu^*$. In the strong-coupling limit the decrease is not given by a difference, but rather is described by the ratio (2.60a). This gives a stronger dependence on μ^*. If the effect of the Coulomb interaction were described by the difference $\lambda - \mu^*$, as in the case of weak coupling, this effect would be negligibly small in the limit of $\lambda \gg 1$. We see that an increase in the electron–phonon coupling is accompanied by a transition to the dependence (2.60a), which is more drastic in the $\lambda \gg 1$ limit than a simple subtraction. As a result, the Coulomb term still makes a noticeable contribution despite the fact that $\lambda \gg 1$.

This change in the dependence on T_c on μ^* is connected with the features of the kernel of Eliashberg equation. In the weak coupling approximation the temperature dependence of the phonon Green's function is negligibly small; this smallness is due to the presence of the small parameter $(T_c/\Omega_D)^2$. In this case the kernel contains the difference $\lambda - \mu^*$.

In the limit $\lambda \gg 1$, on the other hand, the phonon Green's function depends strongly upon T, and this dependence results in a different relation between λ and μ^*.

Therefore, in the limit of very strong coupling T_c is described by the expressions (2.58) and (2.60) which are entirely different from the exponential dependences of the weak-coupling case [see Eqs. (2.46) and (2.50)].

2.3.5. General case

In previous sections we have described several special cases applying to various strengths of the electron–phonon coupling. We have seen the interesting fact that the analytical expressions for T_c are very different for different coupling strengths. The weak and intermediate coupling cases are described by the exponential dependence (2.46) and (2.50), whereas the superstrong coupling case is characterized by the dependence $T_c \propto \lambda^{1/2}$, Eq. (2.58).

Clearly, it is attractive to try to obtain an expression valid for an arbitrary λ. Furthermore, from the practical point of view, it is important to be able to analyze the case of $1.5 \lesssim \lambda \lesssim 5$, which is outside the limits of applicability of both the McMillan equation ($\lambda \lesssim 1.5$) and the very strong coupling limit, Eq. (2.58) ($\lambda \gtrsim 5$). Such a universal equation has been derived in [34].

We will make use of the matrix representation (2.56). The most interesting case corresponds to $v^2 = (\tilde{\Omega}/2\pi T_c)^2 < 1$, with $\tilde{\Omega} = \langle \Omega^2 \rangle^{1/2}$. The region of $v^2 \gtrsim 1$ can be treated analytically and is described by the expression (2.50).

The problem reduces to that of solving the matrix equation $\det |\tilde{K} - \tilde{I}| = 0$. In the region $v^2 < 1$ the convergence is relatively rapid, and to a 1% accuracy it is sufficient to stop at $m = 5$. The calculation can be performed for a range of values of v, and the corresponding values of α determined. As a result, we obtain a dependence of $T_c/\tilde{\Omega}$ upon λ as shown in Fig. 2.3a.

With high accuracy, this curve can be described by the following simple analytical expression:

$$T_c = 0.25\tilde{\Omega}(e^{2/\lambda} - 1)^{-1/2}. \tag{2.61}$$

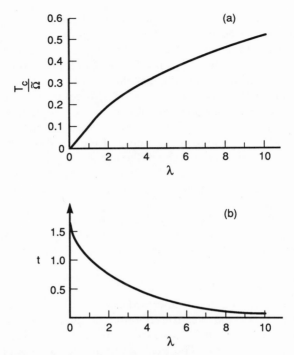

FIG. 2.3. (a) The dependence $T_c(\lambda)$. (b) Universal function $t(\lambda)$.

The function (2.61) can now be used as a trial function in Eq. (2.56) and can be seen to satisfy it with high precision.

Let us consider Eq. (2.61) in some detail. First of all, we can analyze various limiting cases. If $\lambda \gg 1$, the exponential factor can be expanded in powers of λ^{-1} and we find $T_c \cong 0.18\lambda^{1/2}\tilde{\Omega}$ in accordance with Eq. (2.58). In the opposite limit of weak and intermediate coupling ($\lambda \lesssim 1$), we recover an expression essentially identical to Eq. (2.50).

It should be stressed that Eq. (2.61) is not an interpolation. It has been obtained directly from the Eliashberg equation, and explicitly describes T_c for an arbitrary value λ.

As noted above, the general derivation of Eq. (2.61) is described in [32]. Later, it was also recovered in [35] by means of the 1×1 matrix approach (2.56) applied to a very special case: $\alpha^2(\Omega)F(\Omega) = $ const for $0 < \Omega < \Omega_c$ and $\alpha^2(\Omega)F(\Omega) = 0$ for $\Omega > \Omega_c$. This case corresponds to the superstrong coupling limit $\lambda = \infty$, as can be seen from Eq. (2.40): the integrand diverges as $\Omega \to 0$. In this extreme limit the 1×1 approximation used in [35] is justified.

One can see from Table 2.1 that there is a good agreement between Eq. (2.62) and experimental data.

Table 2.1. The ratio $T_c/\tilde{\Omega}$

Superconductor	$[T_c/\tilde{\Omega}]_{th}$	$[T_c/\tilde{\Omega}]_{exp}$
In	0.04	0.04
$Pb_{0.4}Tl_{0.6}$	0.07	0.07
In_2Bi	0.08	0.09
Hg	0.09	0.11
Pb	0.10	0.105
$Pb_{0.9}Bi_{0.1}$	0.125	0.13
$Pb_{0.45}Bi_{0.55}$	0.16	0.15

Let us now consider the effect of the Coulomb interaction on T_c. This question can be analyzed by analogy with the derivation of Eq. (2.61). We need to solve the equation $\det |\tilde{M}^c - \tilde{I}| = 0$. Usually $\mu^* \cong 0.1$, i.e., $\mu^* \ll 1$. In [32] a calculation was carried out for $\mu^* \lesssim 0.2$. The dependence obtained can be described by the expression

$$T_c = 0.25\tilde{\Omega}(e^{2/\lambda_{eff}} - 1)^{-1/2}, \tag{2.62}$$

where

$$\lambda_{eff} = (\lambda - \mu^*)[1 + 2\mu^* + \lambda\mu^* t(\lambda)]^{-1}. \tag{2.62a}$$

The universal function $t(\lambda)$ is presented in Fig. 2.3b. Thus T_c is described by the analytical expression (2.62) which has the same form as Eq. (2.61), with the replacement $\lambda \to \lambda_{eff}$.

We see from Eq. (2.63) that the Coulomb contribution depends on the value of λ. In the strong-coupling limit ($\lambda \gg 1$) we find that $\lambda_{eff} = \lambda(1 + 2.6\mu^*)^{-1}$, in accordance with Eq. (2.60a); in the opposite case ($\lambda < 1$) we find $\lambda_{eff} = \lambda(1 + 2\mu^* + 0.75\lambda\mu^* + 0.8\rho\mu^*)^{-1}$, in agreement with Eq. (2.54). It is interesting to trace the variation of the dependence of λ_{eff} on μ^*. In the limit of weak coupling, the decrease in λ due to μ^* is given by simple subtraction, whereas for superstrong coupling it is described by the relation (2.60a).

On the basis of the McMillan representation (2.52),

$$\lambda = \eta'\tilde{\Omega}^{-2}, \tag{2.63}$$

where $\eta' = \eta/M$, $\eta = \langle I^2 \rangle v(0)$, one can study the dependence of T_c on $\langle \Omega^2 \rangle$. It is of interest to study the direct analytical dependence of T_c on various parameters. As described earlier (see Section 2.3.2), from the McMillan expression one may conclude that there exists a maximum possible T_c at $\lambda \cong 2$. But this result is not convincing because the McMillan equation is not valid for such large values of λ.

Let us now reconsider this question making use of the universal equation (2.61). Based on this result and on Eq. (2.63), we find that $\partial T_c/\partial\tilde{\Omega} > 0$ for any λ. Hence phonon softening always results in an increase in T_c. Note that this increase slows down at large λ, and T_c saturates at the value $T_c = 0.18\eta'^{1/2}$ [see Eq. (2.58)].

2.4. Properties of superconductors with strong coupling

As has been stressed earlier, the weak-coupling approximation (BCS model) is remarkable for the universality of its results. The best-known one relates the energy gap at $T = 0$ K and T_c:

$$2\varepsilon(0) = 3.52T_c. \tag{2.64}$$

The model yields many other universal relations such as $\varepsilon(T)/T_c = 3.06[1 - T/T_c]^{1/2}_{|T \to T_c}$, $\Delta C/C_n(T_c) = 1.4$, etc. The origin of this universality is in the fact that all superconducting properties in the BCS model are determined by the single parameter λ, the coupling constant (or, more rigorously, by $\lambda - \mu^*$). Instead of λ, one can equally well select T_c as such a parameter: it is actually much better to deal with an experimentally measured quantity. As a result, all the other parameters (energy gap, heat capacity, critical field, etc.) are related in a universal way to T_c.

Strong coupling effects destroy this universality. This is caused by the fact that strong coupling theory contains a direct dependence on the structure of the phonon spectrum. It turns out (see below) that an increase in the coupling strength leads to the coefficient $\alpha = 2\varepsilon(0)/T_c$ growing beyond the BCS value of 3.52. Since the ratio α can be measured experimentally, this provides a relatively simple test of the magnitude of the coupling constant. One has to keep in mind, though, that the above argument is valid only if we are dealing with only one gap (and consequently only one coupling constant). This is the case for practically all conventional superconductors. The case of multiband structure requires a separate analysis (see Chapter 6).

Let us look at the relationship between $\varepsilon(0)$ and T_c [36]. We start with the Eliashberg equation, writing it down initially for the case of $T = 0$ K:

$$\Delta(\omega_n)Z = \pi T \sum_{n'} \int d\Omega \frac{g(\Omega)}{\Omega} D(\omega_n - \omega_{n'}, \Omega) \frac{\Delta(\omega_{n'})}{[\omega_{n'}^2 + \Delta^2(\omega_{n'})]^{1/2}} \tag{2.65}$$

(see Eq. (2.2)). This is a nonlinear equation. Consider a model with just one phonon frequency $\tilde{\Omega}$, i.e., let the phonon density of states look like a delta function. This approximation is justified, because $F(\Omega)$ usually contains sharp peaks (see discussion below, p. 32). We will assume that $T_c \ll \tilde{\Omega}$; nevertheless, our goal is to calculate corrections $\sim (T_c/\tilde{\Omega})^2$.

The solution can be found by quasi-linearizing Eq. (2.65). The idea of this method [37] (see also [38]), is as follows. The equation for $\Delta(\omega)$ at

$T = 0$ K can be written in the form

$$V(\omega) = \frac{\tilde{\Omega}^2}{\tilde{\Omega}^2 + \omega_n^2} + Z^{-1} \int_{-\infty}^{\infty} d\omega' \, R(\Omega, \omega, \omega') \frac{V(\omega')}{\sqrt{\omega'^2 + \Delta^2(0)V^2(\omega')}}, \quad (2.65a)$$

where $V(\omega) = \Delta(\omega)/\Delta(0)$ and the kernel is

$$R(\omega, \omega') = \frac{\tilde{\Omega}^2}{\tilde{\Omega}^2 + (\omega - \omega')^2} - \frac{\tilde{\Omega}^2}{\tilde{\Omega}^2 + \omega^2} \frac{\tilde{\Omega}^2}{\tilde{\Omega}^2 + \omega'^2}. \quad (2.65b)$$

It is important that $R(\omega, 0, \Omega) = 0$, and $R(\omega, \omega', \Omega) \sim \omega'^2$ for small ω'. It follows that the main contribution to the integral over ω' comes from the region $\omega' \cong \tilde{\Omega}$, and in the zeroth-order approximation we can neglect the factor $\Delta^2(0)V(\omega')$ in the denominator on the right-hand side of Eq. (2.65a). As a result, we end up with a linear equation for the zeroth-order function $V_0(\omega)$. The solution of the full equation (2.65a) can be sought in the form $V(\omega) = V_0(\omega) + V'(\omega)$, where $V'(\omega) \propto \Delta^2(0)/\tilde{\Omega}^2$. It is also possible to derive a linear integral equation for the function $V'(\omega)$. Thus the quasi-linearization method reduces the solution of the nonlinear equation (2.65a) to that of a system of linear equations.

For $T = T_c$, Eq. (2.65) can be transformed with the aid of the Poisson summation formula:

$$\sum_{\omega_n > 0} f(\omega_n) = (1/2\pi T) \int_0^{\infty} f(z) \, dz$$

$$+ (1/\pi T) \sum_{s=1}^{\infty} (-1)^s \int_0^{\infty} f(z) \cos(sz/T) \, dz. \quad (2.66)$$

By comparing the equations for $\Delta(\omega)$ at $T = 0$ K and $T = T_c$, we can derive an expression relating the energy gap $\varepsilon(0)$ at $T = 0$ K and the critical temperature T_c [26, 36]:

$$\frac{2\Delta(0)}{T_c} = 3.52\left[1 + a\left(\frac{T_c}{\Omega}\right)^2 \ln \frac{\tilde{\Omega}}{T_c}\right], \quad (2.67)$$

where a is a numerical factor ($a \cong 5.3$).

Several comments should be made concerning Eq. (2.67). Clearly, this expression generalizes Eq. (2.64) obtained in the weak-coupling approximation: we recover the latter if the second term in the brackets is neglected. This second term, which reflects the effects of strong coupling, has a peculiar form. In addition to the ordinary quadratic term, it contains the large logarithmic factor $\ln(\tilde{\Omega}/T_c)$.

We see that taking account of the effects of strong coupling destroys the universality of the relation between $\varepsilon(0)$ and T_c. The expression we obtained contains a characteristic phonon frequency which varies from one superconductor to another.

Furthermore, the relationship is no longer that of direct proportionality. This statement can be checked experimentally by studying the dependence of the energy gap and T_c on pressure. Such an experiment [39], which utilized a tunnel junction under pressure, has shown that, indeed, in lead the energy gap and the critical temperature do not vary in an identical manner. This is a manifestation of strong coupling effects. The experimental method employed in this work also allowed control of the shift of the characteristic phonon frequency $\tilde{\Omega}$. The results were in very good agreement with Eq. (2.67).

The following question arises as a result of the foregoing discussion: what should be taken as the characteristic phonon frequency? The expression (2.67) has been obtained using a model with a delta function-like phonon peak; it would clearly be unreasonable to expect precise quantitative agreement if it were applied to real superconductors.

If the phonon density of states consists of two peaks (this case usually corresponds to materials with a simple crystal structure and no optical phonon branches, see, e.g., Fig. 3.5), then we can take the frequency of the lower peak, Ω_1, as the characteristic phonon frequency. Indeed, corrections due to the higher-lying peak are $\sim (T_c/\Omega_2)^2$ and are therefore small.

Note that the two peaks we are describing correspond to the longitudinal and transverse branches of the phonon spectrum. The origin of these peaks is as follows. The main contribution to pairing is made by the short-wavelength phonons. This is different from the case encountered in the analysis of transport properties, but one has to remember that pairing is governed by the exchange of virtual, rather than thermal, phonons. In the short-wavelength part of the spectrum, the simple acoustic dependence $\omega_i = u_i q$ (here i denotes different phonon branches) does not hold. The region of Van Hove singularities, where the frequency is only weakly dependent on the momentum, contains the maximum phonon density of states. As a result, this region corresponds to the observed peaks of the function $\alpha^2(\Omega)F(\Omega)$.

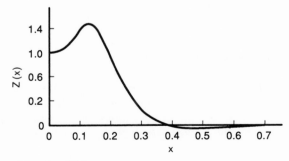

FIG. 2.4. Universal function $Z(x)$.

Instead of Ω_1, we could choose other characteristic frequencies, such as $\langle \Omega^2 \rangle^{1/2}$, $\langle \Omega \rangle$ with the averaging carried out with respect to the function $\alpha^2(\Omega)F(\Omega)$: $\langle \Omega^2 \rangle = \int d\Omega \, \alpha^2(\Omega)F(\Omega)\Omega$, $\langle \Omega_{\log} \rangle$, etc. (see [13, 40]). The comparison between theory and the experimental data depends, of course, on the choice of the characteristic frequency $\tilde{\Omega}$, although the difference is not dramatic. The general qualitative features described above remain valid.

Equation (2.67) shows that one of the effects of strong coupling is to make the ratio $\alpha = 2\varepsilon(0)/T_c$ exceed the value $\alpha_{\text{BCS}} = 3.52$ (the extra term in parentheses is always positive). There is a one-to-one correlation between α and the magnitude of the coupling constant λ. An increase in λ leads to an increase in T_c, and as a result, to an increase in the ratio α.

It is important to keep in mind, however, that, although we were interested in the correction $\propto (T_c/\tilde{\Omega})^2$, the result (2.67) has been obtained under the assumption that $T_c \ll \tilde{\Omega}$. As λ, and consequently T_c, increases, we come to a point (at $\lambda \cong 2$–3) when the above inequality no longer holds. The following question then arises: what happens to the ratio α in the regime of superstrong coupling where the critical temperature is described by Eqs. (2.58) and (2.61)?

It can be shown in general [41] that the ratio $2\varepsilon(0)/T_c$ saturates with increasing λ. Indeed, the energy gap is defined as the root of the equation $\varepsilon = \Delta(-i\varepsilon)$. We are interested in the behavior of the function $\Delta(\omega)$ for high frequencies, where it is determined from the equation that can be obtained from Eq. (2.65)

$$\Delta(\omega) = \frac{\lambda \tilde{\Omega}^2}{\omega^2} \int d\omega' \, \Delta(\omega') |\omega'^2 + \Delta^2(\omega')|^{-1/2}. \tag{2.68}$$

With use of Eqs. (2.65) and (2.68) the expression $\varepsilon(0) = C\lambda^{1/2}$ is obtained for the gap $\varepsilon(0)$ in the regime $\lambda \gg 1$. This dependence is identical with that of T_c [see Eq. (2.58)] in the region of large λ. Consequently, the ratio $2\varepsilon(0)/T_c$ becomes a universal constant as λ becomes large: $\alpha_{\lambda \gg 1} \simeq 12$. Hence, the ratio α has universal values in the regions of both weak and strong coupling, and $\alpha_{\text{BCS}} \equiv \alpha_{\lambda \ll 1} = 1.76 \ll \alpha_{\lambda \gg 1}$. The transition from weak electron–phonon coupling to strong coupling is accompanied by an increase in α, but the dependence $\alpha(\lambda)$ saturates for large λ, even though both T_c and $\varepsilon(0)$ continue to grow.

The fact that $\alpha(\lambda)$ saturates was also deduced from numerical calculations utilizing the delta-function phonon frequency model [40, 42]. This work obtained a numerical value close to that in [41].

The temperature dependence of the energy gap is also different from that in superconductors with weak coupling. It has the form [43]

$$\Delta(T) = a[1 - (T/T_c)]^{1/2}|_{T \to T_c},$$
$$a = 3.06[1 + 8.8(T_c^2/\tilde{\Omega}^2) \ln(\tilde{\Omega}/T_c)]. \tag{2.69}$$

For example, for lead $a \cong 4$. The dependence of $\varepsilon(T)$ influences many properties of superconductors, such as heat capacity, thermal conductivity, ultrasound attenuation, etc. For example, the jump in the heat capacity β is directly related to the change in entropy $S^s - S^n|_{T_0} \propto a^2$.

A more detailed evaluation based on the general expression for the thermodynamic potential [44] leads to the result [43]

$$\beta = 1.43\left[1 + b\left(\frac{T_c}{\Omega}\right)^2\left(\ln\frac{\tilde{\Omega}}{T_c} + \frac{1}{2}\right)\right], \qquad b \simeq 18. \qquad (2.70)$$

For example, for lead $\beta = 2.4$ (cf. $\beta = 1.4$ in the BCS theory) and for gallium $\beta = 2.3$. Ultrasound attenuation as well as the electronic thermal conductivity decrease much sharper with decreasing temperature below T_c than in weakly coupled superconductors [45].

According to the BCS model, the function $H_c/H_{c0} - [1 - (T/T_c)^2]$ is always negative (H_c is the thermodynamic critical field). One can show [17, 43] that strong coupling reverses this sign.

The properties of strongly coupled superconductors can be analyzed in detail by tunneling spectroscopy [17, 46]. This important aspect will be discussed in Chapter 3.

2.5. Electron–phonon interaction and renormalization of normal parameters

The electron–phonon interaction plays an extremely important role in the physics of metals. For example, this interaction is one of the principal relaxation mechanisms defining the electrical conductivity. In this type of situation we are dealing with electron scattering by real thermal phonons. But even in the absence of thermal phonons, e.g., at $T = 0$ K, the polarizability of the lattice (i.e., changes in the regime of zero-point vibrations of the ions) often plays a significant role. Superconductivity represents probably the most spectacular manifestation of this type of interaction. However, they show up in the normal state properties as well. For example, the effective mass m^* measured in certain experiments (e.g., cyclotron resonance or the de Haas–van Alphen effect) comes out different from the so-called band value m^b corresponding to a frozen lattice.

Qualitatively speaking, as a result of lattice polarization electrons find themselves "dressed" in ionic "clouds". This mass renormalization effect turns out to be quite significant for a number of metals. Namely, it is found that $m^*(0) = m^b(1 + \lambda)$ where λ is the coupling constant describing the electron– phonon interaction ($m^*(0)$ is the value of the effective mass at $T = 0$ K). For example, in lead $\lambda \cong 1.4$, so that renormalization leads to the electron becoming about 2.5 times heavier than would be expected for a frozen lattice. A similar relationship holds for the Sommerfeld constant describing the electronic heat capacity: $\gamma(0) = \gamma^b(1 + \lambda)$.

In quantum language, renormalization effects are described by the diagram shown in Fig. 2.1. We are dealing with emission and subsequent absorption of a virtual phonon.

It should be pointed out that some phenomena do not involve mass renormalization (e.g., high-frequency effects, the Pauli spin susceptibility, etc.). A detailed review can be found in the book [5].

Let us come back to the renormalization of the Sommerfeld constant [5, 47, 48]. The entropy of the electron system can be written as [44]

$$S_c = \frac{v_0}{T^2} \int_0^\infty \frac{d\varepsilon\, \varepsilon}{\cosh^2(\varepsilon/2T)} \, [\varepsilon - f(\varepsilon)]. \tag{2.71}$$

Here v_0 is the unrenormalized density of states at the Fermi level and $f(\varepsilon)$ is the odd part of the self-energy function which describes the electron–phonon interaction (see Fig. 2.1).

If we neglect the function $f(\varepsilon)$ on the right-hand side of Eq. (2.71), we recover the usual expressions for the entropy and, consequently, for the heat capacity of a free-electron gas. Renormalization effects are contained in the function $f(\varepsilon)$.

A calculation of the function $f(\varepsilon)$ leads to the following result for the electronic heat capacity:

$$C_e(T) = \gamma(T)T, \tag{2.72}$$

where

$$\gamma(T) = \gamma^0 \left[1 + 2 \int_0^\infty \frac{d\Omega}{\Omega} g(\Omega) Z(T/\Omega) \right]. \tag{2.72a}$$

In this expression $g(\Omega) = \alpha^2(\Omega)F(\Omega)$ [see Eq. (2.37)], and $Z(x)$ is a universal function (Fig. 2.4; do not confuse $Z(x)$ with the renormalization function).

For $T = 0$ K we find

$$\gamma(0) = \gamma^b(1 + \lambda) \tag{2.73}$$

where λ is the electron–phonon coupling constant [see Eq. (2.40)].

As can be seen from Eq. (2.72), the electron–phonon interaction leads to a deviation from the linear temperature dependence of the electron heat capacity. This effect is due to the fact that the state of the lattice changes with temperature as thermal phonons appear.

The result (2.72) can be written in a form which contains only experimentally measurable quantities:

$$\gamma(T) = \gamma(0) \left[1 + \rho \left(\frac{\kappa(T)}{\kappa(0)} - 1 \right) \right]$$

$$\kappa(T) = 2 \int d\Omega\, g(\Omega)\Omega^{-1} Z(T/\Omega) \tag{2.74}$$

where $\gamma(0) = m^* p_F/3$, $m^* = m^b(1 + \lambda)$ is the renormalized effective mass, $\rho = \lambda(1 + \lambda)^{-1}$ is the renormalized coupling constant, and λ and κ are defined by Eqs. (2.40) and (2.72a), respectively.

For $T \to 0$ K,

$$[\gamma(T)/\gamma(0) - 1] \sim T^2 \ln(\Omega_D/T). \qquad (2.75)$$

This dependence was derived in [44]. It is interesting that at sufficiently low temperatures the additional contribution to the electronic heat capacity given, according to (2.75), by $\Delta C \propto T^3 \ln(\Omega_D/T)$ exceeds the lattice heat capacity ($\propto T^3$). As the temperature increases, $\gamma(T)$ goes through a maximum; upon further rise in temperature the function $Z(T/\Omega)$ decreases and approaches zero. It becomes small for $x \cong 0.3$, whereby $\gamma(T)$ acquires the unrenormalized value γ^b. Qualitatively, this is explained by the fact that intense thermal motion washes out the ionic "coat", so that the electrons find themselves "undressed" at high temperatures.

So far we have been discussing corrections to the electronic parameters induced by the electron–phonon interaction. As we have seen, nonadiabaticity leads to significant changes in the effective mass, the electronic heat capacity, the Fermi velocity $v_F = p_F/m^*$, etc. Let us now look at the effect of this interaction on the phonon spectrum. The results here turn out to be completely different [9, 49]. The correction to the phonon frequency $\Delta\Omega$ can be evaluated as the variation of the nonadiabatic contribution to the total energy. A calculation of this correction leads to $\Delta\Omega \sim \kappa^2$; thus the shift in the phonon frequency is relatively small.

This is an important result which has direct bearing on the problem of lattice instability. It also has an interesting history. The fact of the matter is that if one directly uses the Fröhlich model in which the full Hamiltonian is made up of an electronic term, a phonon term with an acoustic dispersion law ($\omega = u_i q$, $q \to 0$), and an interaction term given by H_1 [see Eq. (2.26a)], then one obtains the result $\Omega = \Omega_0(1 - 2\lambda)^{1/2}$.

In other words, one finds a significant shift in the phonon frequency. Furthermore, for $\lambda > \lambda_{max} = 1/2$ the frequency is imaginary, i.e., the lattice becomes unstable. Thus in the early years following the BCS work it was assumed that λ cannot exceed λ_{max}, which then leads to an upper limit on the superconducting critical temperature $T_{c,max} = \Omega_D \exp(-1/\lambda_{max}) \cong 0.1\Omega_D$ achievable by the phonon mechanism. This result is in contradiction to subsequent experimental observations; at present we know many super-conductors with $\lambda > 0.5$.

The problem with the above analysis is that the Fröhlich Hamiltonian is not a valid tool. If we start with the rigorous adiabatic theory, the interaction is given not by H_1 but by the full expression H', see Eq. (2.26). The use of this expression leads to the aforementioned weak renormalization of the phonon frequency $\sim \kappa^2$. A rigorous systematic application of the adiabatic theory leads to only weak corrections due to nonadiabaticity.

2.6. Nonlinear electron–phonon interactions

2.6.1. *Phonon dynamics of perovskites*

Many nonconducting perovskites show a variety of structural transitions corresponding to soft phonon modes, in many cases associated with ferro-and/or antiferroelectric phase transitions [50]. In fact, because of the proximity to a large number of structural transitions and the existence of soft optical modes, speculation has abounded for a good thirty years that ferroelectricity and superconductivity may occur in the same or structurally very similar materials.

It is instructive to consider some aspects of those lattice modes going soft as they seem to behave similarly in both insulating and metallic perovskites. This is strongly suggestive of the fact that the ionic contribution is dominant even after doping with carriers and that the conduction band affects these phonons to a small degree. For example, the structural transition at a temperature of 105 K in $SrTiO_3$ is driven by a rotational displacement of the corner-linked SrO_6 octahedra with a wave vector $q_s = (\pi/a, \pi/a, 0)$ leading to a doubling of the unit cell and an alternation of oppositely tilted CuO_6 octahedra. The softening of this rotational zone-boundary tilt mode as a function of temperature was studied extensively by neutron scattering [51, 52].

It was unexpected in the doped $La_{2-x}Sr_xCuO_4$ material, which for $x > 0.05$ is conducting, that neutron scattering [53] also showed that the tetragonal–orthorhombic transition to be driven by a zone-boundary soft tilt mode of the CuO_6 units of this high T_c superconductor discovered by Bednorz and Mueller [54].

It was furthermore discovered by Axe et al. [55] in $La_{2-x}Ba_xCuO_4$ that a second structural transformation occurs around $T = 60$ K for $x_{cr} = 0.12$. The symmetry changed from the low-temperature orthorhombic (LTO) to a new low-temperature tetragonal (LTT) phase with space group $P4_2/mn$. It was found by Axe that T_c was strongly suppressed in the LTT phase, as discovered earlier by Moodenbaugh et al. [56] and Suzuki and Fujita [57]. In addition, it was found by Crawford et al. [58] that a very large isotope shift is obtained at $x = 0.12$.

A parallel may be drawn between these rigid-body-like motions of subunits of the high T_c copper oxides and organic superconductors, in which the very nature of molecular structure leads to new orientational lattice modes—librons—, which hybridize with the center-of-mass displacements—longitudinal and transverse phonons [59].

It was also found in some theoretical studies that the oxygen motion in certain directions could not be described by harmonic potentials, but led to a double-well potential [60] for precisely those modes which were found to soften and go to zero in the neutron scattering experiments.

Additional indications for unusual phonon dynamics were found in pulsed neutron experiments [61] sensitive to short-range order, extended X-ray absorption fine structure (EXAFS) [62] and ion channelling [63]. Neutron scattering experiments probing the phonon dispersion relations over the entire Brillouin zone performed by the Karlsruhe group [64] found strong indications of multiphonon contributions and additional scattering intensity from additional modes not accounted for by lattice dynamics [65].

It has been proposed by several workers that a double-well potential be used as a starting point for some of the oxygen dynamics [66–70]. Theoretical support for such a potential comes from the nonlinear polarization of the oxygen ion electronic shell [70] in the crystal field of the perovskite lattice and also from LDA calculations [60]. There are two important consequences of such a potential: (1) effects on the phonon dynamics; (2) effects on the carrier–phonon interaction.

To deal with (1), we note that there is extensive literature describing the self-consistent phonon-approximation [71] in ferro- and antiferro electric systems. These treatments have been used in the recent work by Bussmann-Holder *et al.* [67–70]. The second consequence (2) above leads to the appearance of nonlinear terms in the carrier-phonon interaction. A very general study of such higher-order effects was made a long time ago by Geilikman [9].

A particular example of a nonlinear electron–phonon interaction was proposed by Ngai [72] to explain the superconductivity in an early low-transition-temperature perovskite [73]. We have studied a similar model, namely an electron–two-phonon interaction [74] and calculated the potential increase in transition temperature due to this additional interaction.

It is interesting to realize that the great deal of work relating inelastic tunnelling data to evaluation of λ and T_c utilizing the McMillan–Rowell procedure [17, 46, 75, 76] remains within the constraints of linear electron–phonon interactions based on the usual adiabatic approximation. Their success, therefore, may make one pause before proposing additional effects of the type we have in mind.

On the other hand, despite the great variety of traditional super-conductors, it may well be that apart from the BaPbBiO type compounds discovered earlier [77], the perovskites have qualitatively new aspects in their lattice dynamics which suggest that effects such as nonlinear electron–phonon contributions should be studied in more detail.

One of the few studies on the effects of higher-order anharmonicity [78] showed that a repulsive higher-order term (proportional to Q^4) had a very small effect on the order of a few percent on the value of the transition temperature. On the other hand, it was recently argued [79, 80], that a double-well potential would have potentially very large effects on λ due to two effects: The small energy denominators arising from the tunnel-split energy eigenvalues and off-diagonal matrix elements from the anharmonic interactions. Such an approach, however, is outside the well-established Eliashberg equation approach based on harmonic phonon Greens functions.

2.6.2. *Anharmonicity*

We mentioned above that anharmonicity of the lattice should be taken into account. The effect of anharmonicity on the carrier–lattice coupling and T_c is an interesting problem. Important aspects of this problem have studied in [9, 10].

The most convenient framework for the discussion of higher-order electron–phonon interactions is the Eliashberg equations (see Section 2.2) for the electron self-energy $\Lambda(\omega)$ and the superconducting gap function $\Delta(\omega)$. The standard BCS theory is described by the quasi-particle Green's function $G_p(\omega)$, the phonon Green's function $D(q, \omega)$, and Gorkov–Nambu anomalous Green's functions $\Delta(\omega)$ in the superconducting state.

We propose [81] to study the possibility that higher-order electron–phonon interactions may contribute to the gap formation and hence raise the superconducting transition temperature T_c. To simplify the calculation we represent the phonon degrees of freedom by a single optical phonon branch and introduce the following Hamiltonian:

$$
\begin{aligned}
H &= H_0^{el} + H_0^{ph} + H_1, \\
H_0^{el} &= \sum_{i,j} t_{ij}(a_i^+ a_j + \text{h.c.}), \\
H_0^{ph} &= \hbar\Omega \sum_i a_i^+ a_i, \\
H_T &= g_i \sum_{i,j} a_i^+ a_j(b_i^+ + b_i) + g_2 \sum_{i,j} a_i^+ a_j(b_i^+ + b_i)^2.
\end{aligned}
\tag{2.76}
$$

We assume that the electron creation and destruction c_i, c_c^+ describe the conduction band and the phonon creation and destruction operators are a_i^+, a_i. We have written the standard first-order interaction term with coupling constant g_1 and kept the next term in a Taylor series expansion of the electronic energy with respect to the optical phonon coordinate Q.

The coupled Eliashberg equations (EE) for this case have the form

$$
\begin{aligned}
\Delta(\omega_n)Z = {}& \lambda_1\pi T \sum_{m=-\infty}^{\infty} \frac{\Omega^2}{\Omega^2 + (\omega_n - \omega_m)^2} \left.\frac{\Delta(\omega_m)}{|\omega_m|}\right|_{T_c} \\
&+ \lambda_2\pi T \sum_{m=-\infty}^{\infty} \sum_{l=-\infty}^{\infty} \frac{\Omega^2}{\Omega^2 + \omega_l^2} \frac{\Omega^2}{\Omega^2 + (\omega_n - \omega_m - \omega_l)^2} \left.\frac{\Delta(\omega_2)}{|\omega_m|}\right|_{T_c} \\
Z = {}& 1 + \frac{\pi T\lambda_1}{\omega_m} \sum_{m=-\infty}^{\infty} \frac{\Omega^2}{\Omega^2 + (\omega_n - \omega_m)^2} \left.\frac{\omega_m}{|\omega_m|}\right|_{T_c} \\
&+ \frac{\pi T\lambda_2}{\omega_n} \sum_{m=-\infty}^{\infty} \sum_{l=-\infty}^{\infty} \frac{\Omega^2}{\Omega^2 + \omega^2} \frac{\Omega^2}{\Omega^2 + (\omega_n - \omega_n - \omega_l)^2} \left.\frac{\omega_m}{|\omega_m|}\right|_{T_c}
\end{aligned}
\tag{2.77}
$$

(a)   (b)

FIG. 2.5.   (a) One-phonon self-energy part. (b) Two-phonon self-energy part.

Here we have defined Matsubara frequencies for the fermion fields $\omega_n = (2n + l)\pi T$ and $\omega_l = 2l\pi T$ for the Boson field. λ_1 and λ_2 are the one-phonon and two-phonon coupling parameters, defined in terms of the electron–phonon vertices (see Eq. (2.76) and Fig. 2.5) and the one and two-phonon densities of state. Ω_0 is the optical phonon frequency and the EE have been evaluated at the superconducting transition temperature T_c.

The one-phonon MacMillan parameter is

$$\lambda_1 = 2 \int_0^\infty \frac{\alpha^2(\Omega)F(\Omega)\,d\Omega}{\Omega} \tag{2.78}$$

(α is related to the electron–one-phonon matrix element, $F(\Omega) =$ phonon density of states) and we define

$$\lambda_2 = 2 \int_0^\infty \frac{d\Omega}{\Omega} \int_0^\infty \frac{F(\Omega - \Omega')F(\Omega')\beta^2(\Omega')\,d\Omega'}{\Omega'} \tag{2.79}$$

arising from the new electron–two-phonon term in Eq. (2.77).

As we have (over)simplified the phonon system to be described by a single optical (Einstein) mode Ω, $F(\Omega) = \delta(\omega - \Omega)$, and we find trivially

$$\lambda_1 = \alpha^2(\Omega)\Omega^{-1}, \qquad \lambda_2 = \beta^2(2\Omega)\Omega^{-2}. \tag{2.80}$$

We now proceed to solve the coupled EE by following the technique proposed by Owen and Scalapino [33], using the symmetry of the order parameter $\Delta(\omega_n) = \Delta(\omega_{-n})$ and converting to a finite matrix form due to the rapid decrease of the phonon Green's function for n unequal m. Because of the appearance of terms proportional to products of two-phonon Green's functions in Eqs. (2.77), we have a more complicated kernel than in the one-phonon case (Section 2.2.1). After some lengthy but straightforward algebraic steps we obtain the equations in terms of the dimensionless frequency variable $v = \Omega_0/2\pi T_c$:

$$\Delta(\omega_n)Z = \sum_{m \geq 0} [\alpha_1 F_{nm} + \beta_1 G_{nm}] \frac{\Delta(\omega_m)}{|2m + 1|}, \tag{2.81}$$

$$Z(\omega_n) = 1 + \frac{\alpha_1}{2n + 1} \sum_{l=0}^{2m} \frac{1}{v^2 + (n - 1)^2} + \frac{\beta_1}{2n + 1} \sum_{l=0}^{2n} E(l^2) \sum_{m \geq 0} K_{ml}(v^2).$$

To lowest order, we have for the matrices F_{nm} and G_{nm} the values $F_{00} = 1$ and $G_{00} = 16$; $\alpha_1 = \lambda_1 v^2$, $\beta_1 = \lambda_2 v^4$.

This reduces the matrix equations to a quadratic eigenvalue equation for the parameter $v^2 = (\Omega/2\pi T_c)^2$:

$$\lambda_1 v^2 + 16\lambda_2 v^4 = 1, \tag{2.82}$$

and we find the transition temperature T_c to be given by

$$T_c \cong T_c^{(0)}(1 + Z), \qquad Z = \frac{32\lambda_2}{\lambda_1^2}. \tag{2.83}$$

Here $T_c^{(0)}$ is the asymptotic strong-coupling solution discussed in Section 2.3.4 given by

$$T_c^0 \cong \frac{1}{2\pi} \lambda_1^{1/2}\Omega.$$

We have shown how to include the effect of a second-order electron–optical phonon interaction in the Eliashberg equation. By calculating the contribution of such a specific interaction to the transition temperature due to a one-optical-phonon exchange we found that the strong coupling result (Section 2.3.4) can be increased by 10–20%; cf. [82].

We also note that if one takes the presence of strong electron–lattice interactions seriously, as indicated from the various experimental studies mentioned, it may be worthwhile to examine a nonperturbative approach to the strongly interacting electron–phonon problem such as the formation of polarons [83–85] and bipolarons.

2.6.3. Bipolaronic superconductivity and negative Hubbard U-models

A very intriguing possibility based on ideas expressed before the development of the Bardeen–Cooper–Schrieffer (BCS) theory is the notion of "local" electron pairs proposed by Schafroth [86]. In contrast to the BCS picture— which is valid for almost all known superconductors—in which Cooper pair states are made up from time-reversed pair states $(\mathbf{k}, \sigma)$ and $(-\mathbf{k}, -\sigma)$ to form a total $K = 0$ reciprocal space condensate, the "local" pairs in the Schafroth model are paired in configuration space. The discovery in the cuprates of unprecedented short superconducting coherence lengths χ_0 of 20–30 Å provides a compelling argument that the carrier pairing in the cuprates may be in real space [87, 88].

The first discussion of bipolarons—that is, formation of a bound state of two polarons—seems to have been made by Vinetskii [89] and somewhat later by D. M. Eagles [90]. It applies the notion of a polaron studied extensively earlier in condensed matter theory for insulators to two such objects, each carrier strongly dressed by the local polarization field of local atomic displacements and forming a bound state. It can be shown within the Holstein model Hamiltonian [91, 92] which was proposed for such systems that such a state arises for certain values of the electron–phonon coupling constant [89].

One way of formally viewing such a state is based on the model of a negative Hubbard U-term proposed by Anderson [93] and derived from a strong electron–local phonon coupling overcoming the local electron–electron repulsion. Once locally paired, such electronic pairs may hop in their bound state form as suggested in [94, 95]. The analogy to the Hubbard model is a very convenient and powerful one, since many of the known results for magnetic systems may be translated directly with a given key. For example, the superconducting order parameter corresponds to the staggered magnetization and the superconducting transition temperature to the Neel temperature of the corresponding anti-ferromagnet.

The question of the connection between weak coupling BCS coupling and the opposite limit of strongly coupled hard-core bosons (bipolarons) forming a Bose–Einstein condensate was elegantly treated in [96] and shown to occur smoothly.

A detailed model for bipolaronic superconductivity was proposed by Alexandrov and Ranninger [97] to explain the experimental data in the titanium oxides and tungsten bronzes studied by Chakraverty et al. in Grenoble during the middle seventies [98, 99].

With the discovery of the cuprates it was suggested [87, 88] that the high transition temperatures may be due to bipolaronic superconductivity. Subsequently several groups pursued this idea in considerable detail, especially the group in the Grenoble–Moscow–Poznan collaboration of J. Ranninger, A. Alexandrov, J. Micnas and S. Robaszkiewicz [97, 100] and D. Emin [101]. Excellent reviews of these results can be found in [102, 103].

The bipolaronic mechanism of superconductivity assumes the presence of the pre-existing pairs (bipolarons) above the transition temperature, and at $T = T_c$ they undergo Bose–Einstein condensation. The existence of such pairs in the normal state should be verified experimentally. The proponents of the bipolaronic scenario need to assume at least two distinctive components: one fermionic which accounts for the Fermi surface (its existence has been proven by photoemission, positron annihilation, and de Haas–van Alphen data), and another, bosonic (bipolarons). Such a picture runs into difficulties with the nature of the superconducting phase transition at $T = T_c$ or in the strong magnetic field.

2.7. Isotope effect

The isotope effect (1950) played a very important role in the understanding of superconductivity. This effect ($T_c \propto M^{-\alpha}$, $\alpha = 0.5$) reflects the influence of the ionic mass on the transition temperature. Experimentally one can observe considerable deviations from the canonical value $\alpha = 0.5$.

The isotope shift is a complicated phenomenon which is difficult to measure and interpret. Even for a lattice with one atom per unit cell, the scale of the isotope shift can be affected by the Coulomb term μ^* (see Section 2.5.5). The Coulomb pseudopotential depends on the phonon frequency, and this leads to a deviation from the simple value of the isotope shift. If the lattice contains several atoms per unit cell, the picture becomes more complicated [104]. Indeed, even for a two-atomic lattice, one can get any value of α. This can be seen directly from the expression for the two phonon modes [105]:

$$\omega_{ac}^2 = \sum_{\alpha=1}^{3} \gamma_\alpha \left(\frac{1}{M_1} + \frac{1}{M_2} \right)$$
$$- \left[\left(\sum_{\alpha=1}^{3} \gamma_\alpha \right)^2 \left(\frac{1}{M_1} - \frac{1}{M_2} \right)^2 + \frac{4}{M_1 M_2} \left(\sum_{\alpha=1}^{3} \gamma_\alpha \cos q_0 d \right)^2 \right]^{1/2}$$

$$\omega_{op}^2 = \sum_{\alpha=1}^{3} \gamma_\alpha \left(\frac{1}{M_1} + \frac{1}{M_2} \right)$$
$$+ \left[\left(\sum_{\alpha=1}^{3} \gamma_\alpha \right)^2 \left(\frac{1}{M_1} - \frac{1}{M_2} \right)^2 + \frac{4}{M_1 M_2} \left(\sum_{\alpha=1}^{3} \gamma_\alpha \cos q_0 d \right)^2 \right]^{1/2}$$

The substitution $M_1 \to M_1^*$ can lead to any value of α, depending on M_2, force constants, etc. This suggests that such a complex system as LaSrCuO may have a small α, even in the presence of strong electron–phonon coupling.

Anharmonicity also might lead to a drastic effect on the shift. For example, the negative isotope effect in the system PdH with $H \to D$ substitution can be explained by large changes in the elastic constants [106].

3

EXPERIMENTAL METHODS

In this chapter we will describe four of the most important experimental techniques that have been used to probe the most fundamental properties of the superconducting state, the energy gap and the pairing interaction. The techniques that will be described are all spectroscopic, they involve the tunneling of quasiparticles through an insulating barrier, the interaction of electromagnetic waves with a superconducting film or surface, the attenuation of ultrasonic sound waves, and nuclear magnetic resonance.

3.1. Tunneling spectroscopy

The BCS theory was very successful in explaining many of the fundamental properties of the elemental superconductors. However, the mechanism of the pairing in many of the more exotic materials was uncertain. Tunneling spectroscopy has proved to be a very powerful tool both in probing the nature of the interaction as well as verifying the validity of the strong coupling theory of Eliashberg [1]. In this section, we will describe the experimental techniques that are used to obtain the tunneling spectra, present the relevant equations for the tunelling density of states and then describe the beautiful inversion method developed by McMillan and Rowell [2].

3.1.1. *Experimental methods*

There are several experimental methods that have been used to generate tunneling spectra. The most widely used are methods based on the deposition of a barrier and a counterelectrode. The simplest manifestation of this method requires the deposition of the superconducting electrode, the formation of the barrier, either by oxidation of the superconductor or by a deposited insulating layer, and the final deposition of another metallic or superconducting electrode at right angles to the original film. This simple cross stripe structure can be made using contact masks and does not require sophisticated lithography. An example of such a structure is illustrated in Fig. 3.1. More recently, junctions have been made by a trilayer process where the electrodes and the barrier are deposited sequentially in the same deposition run and the junction is defined photolithographically afterwards. This method allows the preparation of small and very uniform tunnel

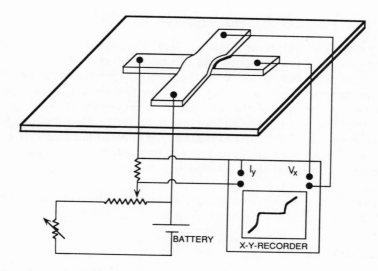

Fig. 3.1. A simple schematic drawing of a cross-stripe junction and the circuitry necessary to measure an $I-V$ characteristic.

structures. Point contacts have also been used very successfully to make tunneling measurements. In this case a sharpened and sometimes oxidized needle with a small voltage applied to it is slowly moved into the proximity of a superconducting surface. The needle position is adjusted until the desired current starts to flow. Current–voltage characteristics can then be measured. This technique has been reasonably successful in probing the oxide super-conductors since it has been very difficult to prepare thin insulating barriers by conventional deposition methods.

The important quantities that need to be measured are the direct current voltage characteristic, $I-V$, the derivative of the $I-V$, dI/dV, and the second derivative d^2I/dV^2. This data must be taken with the sample in both the normal and superconducting state and with the Josephson or pair tunneling contribution quenched by a small magnetic field. These measurements are somewhat tricky and require a.c. lock-in techniques (see [3] and [4]). The $I-V$ characteristic and its first derivative, dI/dV, are essential in determining the tunneling density of states $N_T(\omega)$ and are therefore very important in making the connection to the theory. The first derivative data must be taken to very high precision because only the deviations from the "BCS" (or weak-coupling) density of states are significant and these deviations should be measured to about 1%. Indeed, from these measurements, one can find the energy gap as well as detailed information about the phonon modes responsible for the attractive pairing interaction. Just as importantly, this technique may be unique in determining whether the excitations responsible for the superconductivity are not phononic. Details of these analyses will be described below.

3.1.2. *Energy gap and transition temperature*

In any complete analysis of tunneling data, it is important to know the T_c for the junction. Since the tunneling process involves a very thin region at the surface of the superconductor, the T_c of the junction may be slightly different then the "bulk" T_c for the rest of the film. Since T_c is an important parameter in the rest of the analysis, it is important to know it quite accurately. The simplest method for determining the T_c of the junctions is to plot the ratio of the conductances in the superconducting and normal state for various temperatures and then extrapolate to a ratio of unity. This is the best estimate of the transition temperature of the superconducting material.

Determining the gap from the conductance is simple in principle but in practice is complicated by the nonidealities of real junctions.

3.1.2.1. *MIS junctions*

If the tunnel junction counterelectrode is a normal metal than the normalized conductance of the junction is given simply by

$$\sigma = \left(\frac{dJ}{dV}\right)_s \bigg/ \left(\frac{dJ}{dV}\right)_n = \int N_T(\omega')\left(-\frac{df(\omega' + eV)}{d\omega'}\right)d\omega', \tag{3.1}$$

where $N_T(\omega) = \mathrm{Re}[|\omega|/(\omega^2 - \Delta^2)^{1/2}]$, f is sharply peaked and positive at $E' = -eV$ with a half-width of the order of kT, and Δ is the energy gap. Thus at $T = 0$, $\sigma = N_T(\omega)$. Direct information about the gap is contained in the normalized conductance. In Fig. 3.2 we show the current–voltage characteristics of Al–Pb junctions taken by Nicol et al. [5] above the transition temperature of aluminum. These were the first data where a quantitative comparison between the experiment and the BCS theory were made.

A simple method for determining the gap from arbitrary tunneling data is to measure the normalized conductance σ at zero bias and compare with the tabulated data of σ versus Δ/kT [6]. The complete conductance curve can be fitted to the functional form of Eq. (3.1), which has been fully calculated and tabulated by Bermon [6]. These results are in the weak-coupling limit and they will lose some accuracy as the coupling gets stronger.

3.1.2.2. *SIS junctions*

For junctions with two superconducting electrodes the determination of the gap is more straightforward. If both electrodes are equivalent than for $T \ll T_c$, there should be a sharp discontinuity in the conductance at 2Δ. Of course, for nonideal (real) junctions, the rise is not abrupt but is typically broadened by lifetime effects or by a distribution of gaps. In this case the

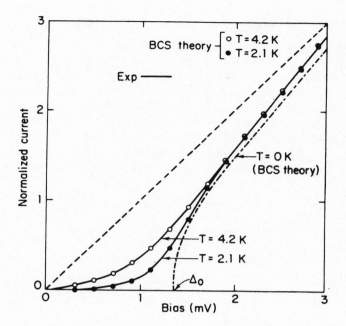

Fig. 3.2. Comparison between experimental and theoretical I–V characteristics of an Al–Pb junction. This figure demonstrates the excellent agreement with the BCS theory. Figure reproduced from [5].

point of maximum slope, dI/dV, should be used. In the case of a linearly linear rise, the midpoint should be used [3].

3.1.2.3. $S_1 I S_2$ junctions

For junctions with nonequivalent superconducting electrodes there is a cusp in the conductance at $\Delta_2 - \Delta_1$ and an abrupt rise in the conductance at $\Delta_1 + \Delta_2$. Thus by clearly locating the position of the cusp and using the same technique for the steep rise that has been described above for the SIS junctions, one can determine the gaps for both superconducting electrodes.

If $T \ll T_c$ then the value of the junction conductivity is an exponentially decreasing function of the smaller gap divided by temperature, and in fact obeys the relation [7]

$$\sigma(V = 0) = \left(\frac{2\pi\Delta_1}{kT}\right)^{1/2} \exp(-\Delta_1/kT). \tag{3.2}$$

Thus the value of the gap for the lower-T_c electrode can be estimated from the temperature dependence of the conductivity at very low bias.

3.1.3. *Inversion of the gap equation and $\alpha^2 F(\Omega)$*

The Eliashberg gap equations describe the direct influence of the phonon density of states, $F(\omega)$ on the energy-dependent gap function $\Delta(\omega)$. In turn the gap function modifies the electronic density of states, which can be directly determined by tunneling conductance measurements. Just how the phonon spectrum influences $\Delta(\omega)$ and the normalized conductance was shown by Scalapino, Schrieffer and Wilkins [8]. They calculated the effect of a single peak in $F(\Omega)$ on $\Delta(\omega)/\Delta_0(\omega)$ and on $N_T(\omega)/N(0)$. The results are shown in Fig. 3.3. It is clear from this calculation that strong peaks in the phonon density of states show up as dips in the normalized tunneling conductance.

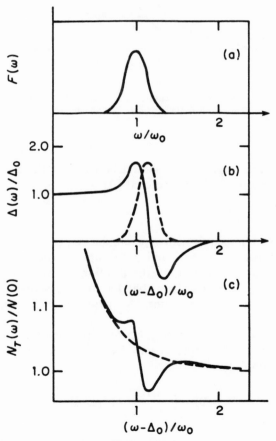

FIG. 3.3. The effect of a single peak in the phonon spectrum (a) on the gap function (b) and on the tunneling density of states. Figure reproduced from [8].

All the information about the phonon density of states and the strength of the electron–phonon interaction are contained in the tunneling measurements. The information is extracted by a numerical inversion of the gap equations.

Thus, inverting the Eliashberg equations to determine the coupled phonon density of states $\alpha^2 F(\Omega)$ and μ^* (the Coulomb pseudopotential) is an extremely powerful method for determining the degree to which any superconductor can be described in the framework of BCS superconductivity, and of course was initially used to determine the accuracy of the Eliashberg theory itself, which was found to be accurate to within several percent. The method developed by McMillan and Rowell [3] involves numerically solving the integral Eliashberg equations for a given set of parameters, calculating the density of states, comparing the calculated values to the measured values, adjusting the input parameters, and iterating the procedure until the calculated density of states matches the measured value. A detailed description of the procedure and the application to lead can be found in [3, 4] but the highlights of this treatment are included here.

The integral equations for the normal and pairing self-energies for a dirty superconductor are

$$\xi(\omega) = [1 - Z(\omega)]\omega = \int_{\Delta_0}^{\infty} d\omega' \, \mathrm{Re}[\omega'/(\omega^2 - \Delta^2)^{1/2}]$$

$$\times \int d\Omega \, \alpha^2(\Omega) F(\Omega)[D_q(\omega' + \omega) - D_q(\omega' - \omega)], \quad (3.3)$$

$$\phi(\omega) = \int_{\Delta_0}^{\Omega} d\omega' \, \mathrm{Re}[\Delta'/(\omega'^2 - \Delta'^2)^{1/2}]$$

$$\times \int d\Omega \, \alpha^2(\Omega) F(\Omega)[D_q(\omega' + \omega) - D_q(\omega' - \omega) - \mu^*], \quad (3.4)$$

where $D_q(\omega) = (\omega + \Omega + i0^+)^{-1}$, $\Delta(\omega) = \phi(\omega)/Z(\omega)$, and $\Delta_0 = \Delta(\Delta_0)$. $F(\Omega)$ is the phonon density of states and $\alpha^2(\Omega)$ is an effective electron–phonon coupling function for phonons of energy Ω.

For an isotropic superconductor the energy-dependent normalized tunneling conductance (see above) gives the superconducting density of states

$$N_T(\omega) = \sigma(\omega) = \mathrm{Re}[\omega/(\omega^2 - \Delta(\omega)^2)^{1/2}]. \quad (3.5)$$

Equations (3.3) and (3.4) are solved by iterating them both together. We start with a guess for $\alpha^2(\Omega) F(\Omega)$ (based on the measured differential conductivity) and μ^*, and a zeroth-order guess for the gap function, i.e., $\Delta^{(0)} = \Delta_0$ for $\omega < \omega_0$ and $\Delta^{(0)} = 0$ for $\omega > \omega_0$ where Δ_0 is the measured energy gap (see above) and ω_0 is the maximum phonon frequency. Using these parameters we find $\zeta^{(1)}, \phi^{(1)}, Z^{(1)}$ and hence $\Delta^{(1)}$. The iteration is continued until $\Delta^{(n)}$ converges to three decimal places. At this point $N_c(\omega)$ is computed and compared to the experimental result, $N_T(\omega)$.

Now the task is to adjust the zeroth-order $\alpha^2 F(\Omega)$ and μ^* to provide convergence of the calculated density of states to the measured density of states. To do this it is necessary to calculate the linear response of the density of states to a small change in $\alpha^2 F(\Omega)$. This amounts to calculating the derivative $\delta N_c(\omega)/\delta\alpha^2 F(\Omega)$. This allows us to estimate the change in the zeroth-order guess for $\alpha^2 F(\Omega)$ necessary for the first-order iteration. The change is given by

$$\delta\alpha^2 F(\Omega) = \int d\omega' \, [\delta N_c(\omega)/\delta\alpha^2 F(\Omega)]^{-1}[N_c(\omega') - N_c^0(\omega')], \qquad (3.6)$$

thus

$$[\alpha^2 F(\Omega)]^{(1)} = [\alpha^2 F(\Omega)]^{(0)} + \delta\alpha^2 F(\Omega)]. \qquad (3.7)$$

Since the gap equations are not linear, the iteration needs to be continued until the calculated density of states reproduces the measured density of states. During the iteration process, μ^* is adjusted so that the calculated energy gap agrees with the measured energy gap. For many conventional superconductors μ^* is found to be approximately 0.1; however, by virtue of the role it plays in the inversion procedure it can be zero or even negative. If some spectral weight is missed in the tunneling data because it occurs at energies beyond the voltage range of the measurements, then the inversion procedure reflects this in a reduced value of μ^*. A recent confirmation of this fact was shown by Schneider et al. [9] for the cluster compound YB_6.

This procedure has been used to show that most of the conventional superconductors are indeed very well described by the Eliashberg equations. For example, the best results have been obtained for lead [4] where the Eliashberg theory has been fully tested and found to be very accurate. Figure 3.4 shows the normalized density of states from the tunneling conductance measurements and Fig. 3.5 shows the calculated $\alpha^2 F(\Omega)$ resulting from the inversion procedure previously described [3, 4].

3.1.4. Electron–phonon coupling parameter λ

The Eliashberg equations can be solved for T_c and the solution can be simplified with the use of the parameter λ which is simply related to $\alpha^2 F(\omega)$ by the following equation:

$$\lambda = \int d\Omega \, 2\alpha^2 F(\Omega)/\Omega. \qquad (3.8)$$

For an isotropic superconductor for any value of λ, Kresin [10] has

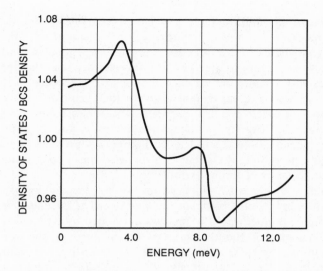

FIG. 3.4. The electronic density of states of Pb divided by the BCS density of states versus $E - \Delta_0$. Figure reproduced from [3].

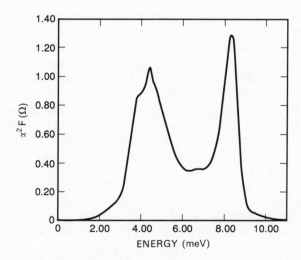

FIG. 3.5. The function $\alpha^2 F(\Omega)$ for Pb found by fitting the data of Fig. 3.4. After [3].

shown that

$$T_c = \frac{0.25\tilde{\Omega}}{(e^{2/\lambda_{\text{eff}}} - 1)^{1/2}}, \tag{3.9}$$

where $\lambda_{\text{eff}} = (\lambda - \mu^*)[1 + 2\mu^* + \lambda\mu^* t(\lambda)]^{-1}$ and $\tilde{\Omega}$ is the average coupled phonon energy. λ is considered to be the most direct measure of the strength of the electron–phonon interaction which strongly effects the nature of the superconductivity; i.e., if $\lambda \ll 1$ we have weak coupling or "BCS" superconductivity, whereas if $\lambda \gg 1$ we have strong-coupled superconductors.

It is also important to note that normal electrons near the Fermi surface are dressed with a cloud of "virtual" phonons. This dressing of the electron properties shows up as an enhancement of the cyclotron mass, the Fermi velocity and the electronic heat capacity. The enhancement over the band value is given by exactly the same parameter λ that is implicit in the Eliashberg gap equations. Thus detailed Fermiology provides an independent method for determining the strength of the electron–phonon interaction as well as the accuracy of the inversion.

In fact, the degree to which phonons are responsible for the superconductivity can be tested by detailed comparisons of calculated specific heat and measured specific heat. The method developed by Kresin [11] depends on measurements of the temperature dependence of the specific heat, determinations of $\alpha^2 F(\Omega)$ from tunneling spectroscopy, and values of $F(\Omega)$ from neutron scattering experiments. The main idea of this procedure is to use the manner in which the phonon dressing of the electronic part of the Sommerfeld constant is unrenormalized at high temperatures. By comparing the high-temperature and low-temperature values of the Sommerfeld constant, one can determine λ, the electron–phonon coupling strength. The electronic specific heat is separated from the total specific heat by integrating the measured $F(\Omega)$ to determine the lattice specific heat and then subtracting it. Tunneling spectroscopy, through its determination of $\alpha^2 F(\Omega)$ is also a measure of λ provided that phonons are fully responsible for the superconductivity. If some other boson contributes strongly to the pairing, then the $\alpha^2 F(\Omega)$ determined by tunneling will not have the correct magnitude since the inversion procedure, which assumes that phonons are fully responsible, will improperly weight the phonon peaks. A simple integration of the tunneling $\alpha^2 F(\Omega)$ will give a value of λ which can be compared to the value determined by the analysis of the specific heat. If they agree, then phonons account for the superconductivity; if they disagree, then some other pairing interaction is present. This procedure has been carried out by Kihlstrom et al. [12] for several A-15 structure superconductors. It was determined that V_3Si was completely phononic, whereas Nb_3Ge was not! It would be very useful to carry out this procedure for the cuprate superconductors. Unfortunately, the state of the art in tunneling spectroscopy at the present time for these compounds is not adequate to provide a very reliable determination of $\alpha^2 F(\Omega)$. Some of the more interesting data on tunneling into the cuprate superconductors will be presented in Chapter 6, which treats the cuprates in great detail.

3.2. Infrared spectroscopy

Study of the absorption of electromagnetic energy has played a very important role in the history of superconductivity. It provided some very early confirmation of the BCS theory and provides a simple method for determining the energy gap especially for thin films where the transmission, the reflection, and the absorption can be estimated. The general equations for the complex conductivity were derived by Mattis and Bardeen [13] and independently by Abrikosov, Gor'kov and Khalatnikov [14]. For the convention where the incoming radiation is in a plane wave $\exp[i(\omega t - \mathbf{qr})]$, then $\sigma = \sigma_1 - i\sigma_2$. The results of the Mattis–Bardeen theory for σ_1/σ_n and σ_2/σ_n are given by the following two equations (following Ginsberg and Hebel [15]):

$$\frac{\sigma_1}{\sigma_2} = \frac{2}{\hbar\omega} \int_\Delta^\infty [f(E) - f(E + \hbar\omega)]g(E)\, dE + \frac{1}{\hbar\omega}$$

$$\times \int_{\Delta - \hbar\omega}^{-\Delta} [1 - 2f(E + \hbar\omega)]g(E)\, dE \tag{3.10}$$

$$\frac{\sigma_2}{\sigma_n} = \frac{1}{\hbar\omega} \int_{\Delta - \hbar\omega, -\Delta}^\Delta \frac{[1 - 2f(E + \hbar\omega)](E^2 + \Delta^2 + \hbar\omega E)}{(\Delta^2 - E^2)^{1/2}[(E + \hbar\omega)^2 - \Delta^2]^{1/2}}\, dE \tag{3.11}$$

If $\hbar\omega > 2\Delta$ then the second term in Eq. (3.10) is included and the lower limit of integration in Eq. (3.11) should be $-\Delta$ instead of $\Delta - \hbar\omega$.

Measuring two of the three rates of energy propagation by reflection, transmission, or absorption for a plane wave normal to the surface of a thin film whose thickness is less than the penetration depth and the coherence length allows a direct determination of σ_1/σ_n and σ_2/σ_n. These results can then be directly compared to the theoretical expressions, Eqs. (3.10) and (3.11). The earliest experiments by Tinkham and Glover [16], however, only measured the transmitted energy, so that unambiguous values of σ_1/σ_n and σ_2/σ_n could not be obtained. However, they could estimate these quantities using the Kramers–Kronig transform relation between σ_1 and σ_2 as well as the sum rule which requires that

$$\int \sigma_{1s}(\omega)\, d\omega = \int \sigma_{1n}(\omega)\, d\omega. \tag{3.12}$$

Equations (3.19) and (3.11) predict that for $T = 0$, σ_1/σ_n is zero for frequencies up to the gap frequency $2\Delta/\hbar$ and finite above this frequency, whereas for finite temperatures σ_1/σ_n will be finite but vanishingly small if $(-\Delta/kT) \ll 1$. The Kramers–Kronig transformation requires that σ_2/σ_n also be small near the gap frequency, which means that both the reflection and absorption will be small and the result is that the transmission has a sharp peak at the gap frequency. Thus a rather unambiguous determination of the gap was made by these early infrared measurements.

Later measurements were made of both transmission and reflection. In these cases, rather accurate determination of σ_1/σ_n and σ_2/σ_n could be

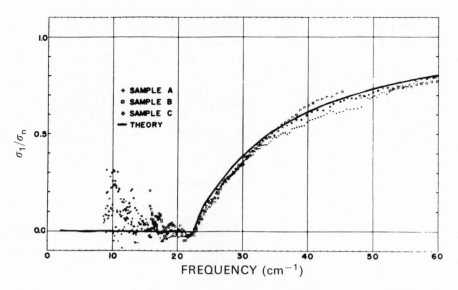

FIG. 3.6. Measured and calculated value of the conductivity ratio for three Pb films as a function of the phonon frequency. Reproduced from [17].

made and compared to the theory Eqs. (3.10) and (3.11). An example of this type of result is shown in Fig. 3.6, which shows measured curves of σ_1/σ_n as well as the theoretical curve based on the Mattis–Bardeen theory [17].

Direct measurements of the absorption of electromagnetic energy as a function of frequency can be used to accurately determine the energy gap. If the temperature is low enough, so that $\exp[-\Delta/kT] \ll 1$, then there is a well defined absorption edge at 2Δ. Measurement of the absorption is done typically by bolometric methods, where the sample is thermally isolated from its surroundings and the absorption is measured by the temperature rise.

3.3. Ultrasonic attenuation

Historically, the first method used to measure the energy gap and check the validity of the temperature dependence of the BCS gap was ultrasonic attenuation. The experimental method involves mounting transducers on opposite sides of a crystal of the superconducting material, launching a sound wave into the sample and measuring the ratio of the attenuation below the transition temperature to its value just above the transition. The simplest case to consider is for the propagation of longitudinal sound waves, where the interaction of the sound wave with the crystal is due to the change in the crystal potential which accompanies the wave. If the phonons associated with the sound wave have energy ($\hbar\omega$) less than the gap energy, 2Δ, then the attenuation at very low temperatures ($T \ll T_c$) should be proportional to $\exp[-\Delta/kT]$ since only the thermally excited quasiparticles can absorb the phonons.

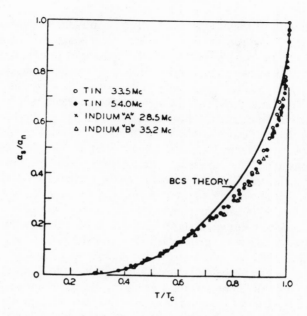

FIG. 3.7. The attenuation ratio as a function of reduced temperature compared to the BCS variation. Reproduced from [20].

Bardeen, Cooper, and Schrieffer in their landmark paper [18] calculated the ratio of the ultrasonic attenuation in the superconducting state, α_s, divided by the attenuation in the normal state, α_n, for the case when $ql \gg 1$ where q is the wavenumber of the sound wave and l is the electron mean free path and $\hbar\omega < 2\Delta$. In this case the ratio α_s/α_n is equal to $2f(\Delta(T), T)$ where f is the Fermi function. Kresin [19] showed that the expression was valid even for $ql \ll 1$. Thus by fitting α_s/α_n as a function of temperature to the Fermi function one can determine the value of the energy gap and its temperature dependence. Figure 3.7 is the result from the paper of Morse and Bohm [20], where the measurements on tin and indium are compared with the BCS theory.

Thus ultrasonic attenuation is clearly an excellent method for determining the energy gap, especially energy gap anisotropy, when these measurements can be made on high-quality samples.

3.4. Nuclear magnetic resonance

Another important probe of the superconducting state, particularly the energy gap, is nuclear magnetic resonance. In these experiments, it is possible to probe the interaction of the magnetic moments of the conduction electrons with the nuclear moments. The magnetic moment of the conduction electrons modifies the static magnetic field at the nucleus. This shift in the resonance field due directly to the susceptibility of the conduction electrons is called the Knight shift. Also the nuclear spins in a metal are in thermal equilibrium with their surroundings by virtue of the spin–spin interactions with the conduction electrons near the Fermi surface. If the temperature or field is changed then the nuclear spin system relaxes to a new equilibrium with a time T_1 that is characteristic of the spin–lattice interaction.

In practice, some unconventional protocols had to be established to measure the magnetic resonance since the field exclusion provided by the Meissner effect for Type I superconductors made the more conventional methods unusable. A method was established by several groups [21] which was able to measure T_1 in the superconducting state in the following way. The nuclear spins were in thermal equilibrium at a magnetic field large enough to quench the superconductivity. Thus a spin temperature (proportional to magnetic field) was established. The field was then quickly turned off, the materials became superconducting and the nuclear spins began to relax to a new effective spin temperature during some variable time t. The field was then re-established in a time short compared to t and the resonance was measured as the field was increasing, capturing the amount of nuclear magnetization left after time t. The size of the resonance peak is measured as a function of t and this gives a direct measurement of the temporal evolution of the nuclear magnetization and hence of T_1.

The details of the evaluation of the temperature dependence of the relaxation rate $(1/T_1)$ in the superconducting state are beyond the scope of this book. However, it is important to point out that crucial role played by the combination of the Fermi function with the superconducting density of states with its concomittant energy gap Δ at the Fermi energy that occurs in the theory. These factors give rise to coherence effects which manifest themselves in different ways depending on the spectroscopy as discussed previously. In the case of NMR relaxation the result is given by the equation

$$R \propto \int_\Delta^\infty dE_i \int_\Delta^\infty dE_f (1 + \Delta^2/E_i E_f) f(E_i)[1 - f(E_f)] N_s(E_i) N_s(E_f) \quad (3.13)$$

Let us analyze this equation in the superconducting state. If we first consider temperatures just below the critical temperature then the width of the factor $f(E)[1 - f(E)]$ is larger than Δ. Thus the fact that the density of states is strongly peaked in the superconducting state and enters the equation for the rate squared combined with the fact that there are many states accessible because of the large relative width of the distribution function $f[1 - f]$ compared to the gap then relaxation can occur faster than in the normal state where the density of states is rather constant. Thus there is an initial rise in the relaxation rate just below T_c. On the other hand, at low temperatures the product $f[1 - f]$ has an appreciable value only within the gap where there are no states. Thus the relaxation is slowed down considerably and in fact depends exponentially on inverse temperature with Δ as the characteristic energy. Thus there is a peak near T_c called the Hebel–Slichter peak [21] followed by an exponential decrease in the relaxation rate. Thus the energy gap can be found directly from the temperature dependence of the relaxation rate well below T_c. The overall temperature dependence has been observed for many superconductors and is a strong consequence of the pairing interaction that is the hallmark of the BCS [18] theory. However, it has recently been shown [22] that for strong electron–phonon coupling there is an additional damping term which arises from the imaginary part of the order parameter which enters the relaxation equation and which, if sufficiently strong, can completely suppress the peak and modify the low temperature behavior as well. This latter result is a consequence of the fact that the presence of the imaginary part of the order parameter is equivalent to the appearance of states in the gap.

The Knight shift which measures the effect of the conduction electrons moment on the nuclear field can also be an important probe of the superconducting state. The Knight shift in the superconducting state can be determined in experiments very similar to those described above for the relaxation rate. In this case, what is of interest is not the amplitude of the resonance signal as a function of t but the field at which it occurs as a function of temperature. This field at which the nuclear resonance occurs is affected by the Pauli susceptibility of the conduction electrons. Well below T_c the susceptibility of the conduction electrons vanishes exponentially with the energy gap. Thus the Knight shift should vanish in the same way. The Yoshida function [23] describes the complete temperature dependence of the Knight shift in the superconducting state.

One of the reasons that NMR can be a unique tool in studying complex superconductors is that the resonance is site and element specific. Thus the local magnetic environment around different nuclei can be probed. If, for example, there are spacially separated superconducting subsystems with differing gaps, they might be able to be separately probed.

4

ELECTRONIC MECHANISMS

Up to this point we have focused on the phonon mechanism of super-conductivity. However, electron pairing can be created not only by phonon exchange but by exchange of other excitations as well. In this chapter we will discuss the electronic mechanism of superconductivity. The most common (but not unique) instance of such a mechanism arises in the presence of two groups of electronic states. Excitations within one of these groups serve as "agents" giving rise to pairing in the other group.

The study of electronic mechanisms of superconductivity began with the work of Little [1]. This work also served to initiate the discussion of the relationship between dimensionality and the superconducting transition. Below we shall describe some relevant models.

It should be be stressed that the phonon and electronic mechanisms are not at all mutually exclusive. It is quite realistic for the two mechanisms to coexist, with both contributing to the superconducting pairing.

4.1. The Little model

Little [1] has considered the case of a one-dimensional organic polymer. Remarkably, at that time no superconducting polymers existed; there were also no organic superconductors. The first superconducting polymer was synthesized in 1975 [2, 3], whereas organic superconductivity was discovered in 1980 [4] and at present represents one of the most actively developing areas of research (see, e.g., [5–7]).

In the Little model, the conduction electrons in the primary chain (see Fig. 4.1) are paired thanks to their interaction with electronic excitations in the side branches. The critical temperature is given by a BCS-like expression (2.46), but in the present case the preexponential factor corresponds to the electronic, rather than ionic, energy scale, so that

$$T_c = \Delta E_{el} \exp(-\lambda_{el}). \tag{4.1}$$

The fact that $E_{el}(\Omega) \gg \hbar\tilde{\Omega}_D$ is responsible for the hope that high critical temperatures may be achieved in this way.

The idea of searching for an electronic mechanism is very fruitful, but its realization is quite nontrivial. Clearly, the preexponential factor acts so as

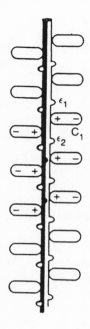

FIG. 4.1. Little's model.

to raise T_c; however, the coupling constant λ also depends sensitively on ΔE_{el}, and it is essential that this constant should not turn out to be too small.

Furthermore, we are really dealing with the difference $\lambda - \mu^*$. Because of the inequality $\Delta E_{el} \gg \hbar\tilde{\Omega}_D$, the logarithmic weakening of the Coulomb repulsion is not as strong as in the phonon case. This factor also decreases the exponent. Finally, certain fundamental problems concerning the superconducting transition arise in low-dimensional cases.

The problem of the intensity of the pairing interaction has been discussed in [8–10]. According to quantum mechanical calculations, in the ordinary model of the type shown in Fig. 4.1 the average spacing between the conducting chain and the charge localized on the side branches is quite large (~ 4.5 Å). As a result, the value of λ is quite low. As a consequence, a different approach was developed in [10]. Here pairing takes place in a chain containing a transition-metal atom (see Fig. 4.2). The conduction electrons move in the $d_{z^2} - p_z$ orbital band. Strong coupling occurs due to the overlap of the d_{xz} and d_{yz} orbitals with those of the ligands. In this situation it is critical that the energy levels of the conduction electron orbitals be close to those of the ligands' electrons.

It is also necessary to have very dense packing of the polarizable ligands. All this works to strengthen the coupling and to optimize the delicate balance of factors controlling the electron–electron interaction.

Another important problem has to do with fluctuations [11–13]. The

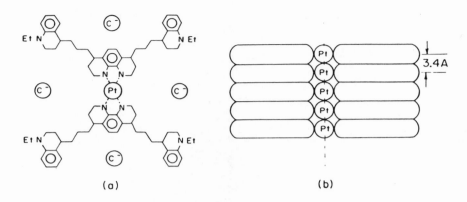

FIG. 4.2. Transition-metal atom chain: (a) top view; (b) side view.

fact of the matter is that fluctuations of the order parameter destroy long-range order: the correlation function $K(x_1 - x_2)$ vanishes for $|x_1 - x_2| \to \infty$ at finite temperatures. This difficulty can be overcome by creating "quasi-one-dimensional" systems made up of bundles of superconducting threads [14]. The probability of jumps between the threads is small but sufficient to stabilize the system with respect to fluctuations.

In the one-dimensional case the Fermi "surface" is made up of two points at $\pm k_F$. This imposes severe restrictions on the allows values of transferred momenta. The allowed processes are shown in Fig. 4.3; each one is characterized by an amplitude g_i. The different possible types of instabilities in the one-dimensional case form the subject of g-ology (see [8]). Specifically, one can introduce various components χ_i of the generalized susceptibility, so that in the random-phase approximation χ_i have the form

$$\chi_i = \frac{\chi_i^0}{1 - \lambda_i \ln(T/D)}. \tag{4.2}$$

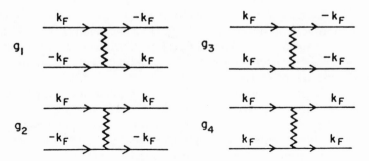

FIG. 4.3. Interaction processes in 1D Fermi system.

Here D is the energy cut-off, and the coupling constants are

$$\lambda_{SS} = 0.5(\tilde{g}_1 + \tilde{g}_2), \qquad \lambda_{TS} = 0.5(\tilde{g}_2 - \tilde{g}_1),$$
$$\lambda_{CDW} = 0.5(2\tilde{g}_1 - \tilde{g}_2), \qquad \lambda_{SDW} = 0.5\tilde{g}_2, \qquad (4.3)$$

where $\tilde{g}_i = g_i/(\pi v_F)$, SS and TS correspond to singlet and triplet super-conducting states, and CDW and SDW denote transitions with the appearance of charge- and spin-density waves, respectively. A rigorous treatment of various instabilities has been carried out in [15, 16].

4.2. "Sandwich" excitonic mechanism

As described above, the main idea of the Little model was to make use of two groups of electronic states. Ginzburg [17, 18] has analyzed the case of two-dimensional conducting layers with coupling due to the exchange of excitons in an adjacent dielectric film. Thus the idea is to create a heterogeneous layered structure with alternating metallic and dielectric planes.

A similar geometry was considered in [19], but with semiconducting rather than dielectric layers. Clearly, the effectiveness of such a system is highly dependent on the thickness of the metallic film, which must be on the order of the screening length, i.e., it may not exceed 10–15 Å.

In the model [19], an important parameter is the penetration depth D of an electron tunneling from the metallic film into the semiconducting covering. This depth can be shown to be equal to

$$D = \frac{\pi}{3} \frac{\hbar^2 \kappa_F}{2m\Delta}. \qquad (4.4)$$

Substituting $\Delta \cong 1$ eV and $\kappa_F \cong 1.5$ Å^{-1}, we find $D \cong 5$ Å. Thus we are dealing with quite substantial penetration depths.

4.3. Excitons and high T_c

The model of excitonic superconductivity has been applied to the high T_c cuprates in [20]. According to this work, the superconducting state in the YBaCuO compound is due to the presence of two subsystems (CuO planes and CuO chains). The chains are treated as an exciton subsystem. The plane–chain interaction leads to pairing of the in-plane conduction electrons. A similar excitonic mechanism was also studied in [21, 22]. Note that the superconducting oxide LaSrCuO does not contain the chains and the explanation of its superconductivity therefore requires a different mechanism.

Another electronic mechanism has been proposed in [23] (see also the review [24]). This version is based on a charge transfer resonance between the $Cu^{3+}O^{2-}$ and $Cu^{2+}O^-$ states. This model does not require the presence of two different subsystems like planes and chains. The transverse resonance corresponds to an energy on the order of 0.5 eV.

Of course, the justification of the excitonic mechanism requires, first of all, their existence in the normal state and this has to be verified experimentally. Secondly, even if the excitons exist in the cuprates, one should prove that they are responsible for the pairing. A large energy scale allows us to use Eq. (4.1) and superconductivity would correspond to the weak coupling case, but nevertheless λ_{ext} should be sufficient. If $\Delta E_{ext} \approx 1$ eV, $T_c \approx 10^2$ K, then λ_{ext} needs to be as 0.2–0.25.

4.4. Three-dimensional systems

A system with two different coexisting electron groups can be realized in three dimensions as well. This case was first studied by Geilikman [25–27]. Here one has to consider two special cases: (1) the group providing attraction is localized, and (2) both groups are delocalized. Let us briefly discuss these two cases.

4.4.1. *Pairing of conduction electrons via interaction with localized states*

For example, we can consider a metal with nonmetallic impurities. The interaction of the conduction electrons with the electrons in the nonmetal atoms leads to an additional effective interaction between the conduction electrons. For clarity, let us write this additional interaction only to the second order in perturbation theory. We obtain

$$\Gamma_{p_1 p_2; p_3 p_4} \approx \Gamma'_{p_1 p_2; p_3 p_4} + \Gamma''_{p_1 p_2; p_3 p_4}, \tag{4.5}$$

where

$$\Gamma'_{p_1 p_2; p_3 p_4} \approx \frac{V_q}{1 - \Pi(\mathbf{q}, \omega)},$$

$$\Gamma''_{p_1 p_2; p_3 p_4} \approx \sum_{n, \lambda, \lambda'} V_{p_1, n\lambda; p_3, n\lambda'} V_{n\lambda', p_2; n\lambda, p_4} \Pi_{\lambda\lambda'},$$

$$V_q = 4\pi e^2/q^2, \qquad \Pi_{\lambda\lambda'} = \frac{n_\lambda - n_{\lambda'}}{E_\lambda - E_{\lambda'} + \hbar\omega},$$

$$\hbar\mathbf{q} = \mathbf{p}_3 - \mathbf{p}_1, \qquad \hbar\omega = \xi_3 - \xi_1,$$

and $\mathbf{p}_i$ are the momenta of the conduction electrons. The first term in (4.5) corresponds to the usual screened Coulomb repulsion, and the second term to the attractive interaction. Here $V_{\mathbf{p}_1 n\lambda; \mathbf{p}_3 n\lambda}$ is the matrix element of the interaction between the delocalized conduction electrons and the localized electrons, λ is the set of quantum numbers of the electron in the nonmetal atom at lattice site n, E_λ is the energy of this electron, and $\Pi(\mathbf{q}, \omega)$ and $\Pi_{\lambda\lambda'}$ are polarization operators. Because of the summation over n, $\Gamma'' \propto N_\alpha$, N_α being the concentration of the nonmetal atoms.

Superconducting pairing can arise if the attraction, given by the second term in (4.5), outweighs the Coulomb repulsion. This requires that the spacing $\Delta E = E_{\text{loc}}$ be sufficiently small. On the other hand, this smallness leads to a decrease in T_c, see Eq. (4.1). Therefore, there exists an optimum value for ΔE. It appears that ΔE should be on the order of 0.1–0.5 eV. Such small values are possible, for example, in the case of fine-structure levels.

A number of factors serve to decrease λ. These include the exchange interaction [28], and the smallness of the ratio a/L (where a is the radius of the orbit of the localized electrons, and L is the lattice parameter). As a result, it is hard to expect this mechanism to provide high values of T_c; still, the study of this mechanism is certainly quite interesting.

It should be pointed out that not only electronic but also vibrational states can be localized. For example, a molecule can be placed on the surface of a thin film or introduced into the sample. Then the interaction of the carriers with molecular vibrations can provide additional interelectron attraction [29].

A special case is that of a superconducting molecular crystal. Systems of this type (for example, the A_3C_{60} systems, where A is an alkali atom) have lattices made up of molecules with their own internal electronic and vibrational levels.

It is interesting to consider the case of pairing provided both by the ordinary phonon mechanism and by an additional interaction with localized electronic states. Making use of the step method, one finds in the weak-coupling approximation:

$$T_c \cong \tilde{\Omega} e^{-1/g_0} \tag{4.6}$$

where

$$g_0 = g_{ph} - (g_e - \bar{g}_c)/[1 - (g_e - \bar{g}_c)\ln(\Delta E_a/\hbar\tilde{\Omega})]$$

Note that the difference $g_e - g_c$ enters the expression for g_0 with a minus sign. This means that the additional attraction is logarithmically enhanced.

4.4.2. Two delocalized groups

A case of this type is realized, for example, in a metal with two overlapping bands of very different widths, e.g., an s- and a p-band. Virtual electronic transitions in the narrow band (denoted b), caused by the Coulomb interaction of the a and b electrons, can lead to pairing in the wide band. A calculation within the random-phase approximation, which is at least of qualitative interest when applied to real metals, leads to the following expression for the effective interaction of the a electrons:

$$\Gamma_{aa} = (V_{aa} + \Pi_{bb}R)S^{-1}, \qquad \Gamma_{bb} = (V_{bb} + \Pi_{aa}R)S^{-1}$$
$$S = 1 + V_{aa}\Pi_a + V_{bb}\Pi_b + \Pi_a\Pi_b R, \qquad R = V_{ab}^2 - V_{aa}V_{bb} \tag{4.7}$$

Here Π_{aa} and Π_{bb} are the polarization operators for the a and b electrons, and V_{ik} are the Coulomb matrix elements. Conditions under which the effective interaction becomes attractive turn out to be rather nontrivial.

The mechanism described here is also possible in multivalley semiconductors, where the large value of the dielectric constant ε makes the random-phase approximation more applicable [30].

Systems with overlapping bands can give rise to low-lying collective branches (acoustic plasmons). The exchange of these collective excitations also can lead to pairing. We will discuss this mechanism in Section 4.6.

4.5. Negative U-centers

Negative U-centers represent a set of localized electronic orbitals. A U-center exhibits on-site attractive interaction due to coupling to some additional degrees of freedom. Originally, local electron–phonon coupling was assumed responsible for this interaction [31, 32]. One can also study coupling to excitonic degrees of freedom [33–35].

The localized electronic states are hybridized with the conduction band. As a result, pairing of the localized states enhances the critical temperature of the entire system.

There is a key difference between the BCS and the usual excitonic mechanisms (Sections 4.1–4.4) and the mechanism described here. In the former case all conduction electrons participate in boson exchange, whereas in the case of U-centers this exchange takes place only for the localized electronic states or, in a more qualitative sense, only for the electrons which tunnel into U-center orbitals. The model Hamiltonian studied in [34, 35] has the following form:

$$\hat{H} = \hat{H}_0 + \hat{H}_i, \tag{4.8}$$

$$H_0 = \sum_{p,s} \varepsilon_p c_{ps}^\dagger c_{ps} + \sum_{p,s} N^{-1/2}(V_p d_s^\dagger c_{ps} + \text{h.c.}), \tag{4.9}$$

and

$$H_i = \varepsilon_d n_d + \lambda \sigma_x(n_d - 1) + \eta \sigma_x + \tfrac{1}{2}\Omega \sigma_z. \tag{4.10}$$

The first term in (4.9) corresponds to the conduction band, and the second term describes its hybridization with the localized orbital d. V_p is the hybridization matrix element, and n is the number of host lattice sites. The $\hat{H}_i$ term describes the localized state, n_d is the on-site occupation number, and ε_d is the energy.

A very important part of the Hamiltonian is the second term in (4.10), which represents coupling to the two-level exciton subsystem described by the Pauli operators (λ is the strength of the coupling, and Ω and η are the exciton frequency and the asymmetry energy, respectively). A detailed analysis with the use of the diagrammatic technique and Monte-Carlo simulations leads to a number of interesting conclusions. The pairing mechanism turns out to be very efficient. For example, if the bandwidth is about 3 eV, the attraction $U = 0.3$ eV, and the impurity concentration $C = 4.5\%$, one can achieve $T_c = 20$ K.

However, note that according to [35] the parameters of the model have to be fine-tuned in order to obtain high T_c; elimination of the relatively narrow resonance, e.g., by changing the stoichiometry, leads to a drastic effect on the critical temperature.

The model of negative U-centers has been applied to the high T_c oxides (see, e.g., [36, 37]) with the oxygen vacancies in the CuO planes associated with the U-centers.

4.6. Plasmons

In this section we focus on a different type of electronic excitations: plasmons, Bose-type collective excitations of the electron gas.

Plasmons describe collective oscillations of the electrons with respect to the positive background. The concept of such excitations for the quantum Fermi-gas was introduced with the use of the quantum kinetic equation in [38], and by the collective variable method in [39].

The spectrum of plasma oscillations is given by the poles of the two-particle Green's function (Fig. 4.4) or by the poles of the vertex Γ. Since the function Γ also enters the equation for pairing, it is clear that the present problem is also relevant to the analysis of the superconducting case.

Consider an isotropic three-dimensional system. Usually the plasma frequency is quite high (~ 5–10 eV). However, when dealing with more complicated band structures one encounters additional low-lying plasmon branches.

4.6.1. Overlapping bands. "Demons"

Consider the case of a metal with two overlapping bands, one of which contains light carriers and the other is narrow and contains heavy carriers. It turns out that its energy spectrum is characterized by the presence of an acoustic plasmon branch ("demons"), introduced in [40]. This mode corresponds to the collective motion of the light carriers with respect to the heavy ones. This acoustic branch is similar to phonons and can probide inter-electron coupling.

Note, by the way, that low-frequency plasma excitations can arise in

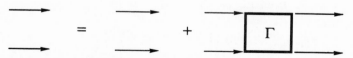

FIG. 4.4. Two-particle Green's function; Γ is a total vertex.

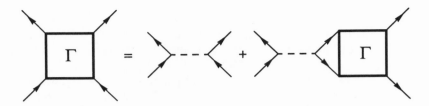

FIG. 4.5. Diagrammatic equation for the total vertex.

low-dimensional systems even in the simplest one-band case. These cases will be considered later (Sections 4.6.2 and Chapter 7).

Let us focus on the 3D case. Consider first the case of a single band. In the random-phase approximation (RPA) the equation for Γ has the following form (see Fig. 4.5):

$$\Gamma(\mathbf{q}, \omega) = V(\mathbf{q}) + V(\mathbf{q})\Pi(\mathbf{q}, \omega)\Gamma(\mathbf{q}, \omega), \qquad (4.11)$$

where $V(\mathbf{q}) = 4\pi e^2/q^2$ is the Fourier component of the Coulomb interaction and

$$\Pi(\mathbf{q}, \omega) = \sum_{\kappa} \frac{n_{\kappa+\mathbf{q}/2} - n_{\kappa-\mathbf{q}/2}}{\omega - \varepsilon_{\kappa+q/2} - \varepsilon_{\kappa-q/2}} \qquad (4.12)$$

is the polarization operator. From Eq. (4.11) we find

$$\Gamma = \frac{V}{1 - V\Pi}. \qquad (4.13)$$

The function $1 - V\Pi$ is the effective dielectric function $\varepsilon(\mathbf{q}, \omega)$, since, according to Eq. (4.13), the effective electron–electron interaction can be written as V/ε.

As mentioned above, plasmons are defined by the poles of Γ or, equivalently, by the zeros of the dielectric function. They are thus the roots of the equation

$$1 - V(\mathbf{q})\Pi(\mathbf{q}, \omega) = 0. \qquad (4.14)$$

Let us introduce the dimensionless parameter $\alpha = \omega/qv_F$. One has to distinguish two cases: $\alpha < 1$ and $\alpha > 1$. Plasmons appear in the region of $\alpha > 1$. For simplicity, consider the case $\alpha \gg 1$. Then the polarization operator can be evaluated from Eq. (4.12) and is given by

$$\Pi(\mathbf{q}, \omega) = v(v_F q/\omega)^2, \qquad (4.15)$$

where v is the density of states.

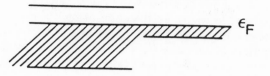

ϵ_F

FIG. 4.6. Overlapping bands.

Making use of this expression, we find from Eqs. (4.14), (4.15), and the expression for $V(q)$ the well-known result for the plasma frequency:

$$\omega_{\mathrm{pl}} = (4\pi e^2 n/m)^{1/2}. \tag{4.15a}$$

In the opposite limit of $\alpha \ll 1$, the polarization operator is given by $\Pi_0 \equiv \Pi_{\alpha \ll 1} = -m k_F/\pi^2$. Then Eq. (4.13) describes screened Coulomb interaction, so that $\Gamma = 4\pi e^2/(q^2 + k_D^2)$ with $k_D^2 = 4\pi e^2 |\Pi_0|$.

Consider now the more interesting case of two overlapping case of two overlapping bands containing light and heavy carriers. One of the bands is broad ("light" carriers), whereas the other one is narrow ("heavy" carriers, see Fig. 4.6). Then instead of (4.13) we have [see Eq. (4.7)]

$$\Gamma_{aa} = (V_{aa} + \Pi_{bb}R)S^{-1},$$
$$S = 1 + V_{aa}\Pi_a + V_{bb}\Pi_b + \Pi_a\Pi_b R, \qquad R = V_{ab}^2 - V_{aa}V_{bb}, \tag{4.16}$$

where V_{aa}, V_{bb}, and V_{ab} are the Coulomb matrix elements. In this case Γ_{aa} is defined by two parameters: $\alpha_a = \omega/v_{Fa}q$ and $\alpha_b = \omega/v_{Fb}q$, where the indices a and b refer to the two bands. Let us assume that $m_b \gg m_a$ and $v_{Fb} \ll v_{Fa}$ and consider the case of $\alpha_a \ll 1$ and $\alpha_b \gg 1$. Then

$$\Pi_{a;0} = -m_a k_F/\pi^2, \qquad \Pi_{b|\alpha_b \gg 1} = (v_{Fb}q/\omega)^2. \tag{4.17}$$

In this case the light carriers provide screening, while the presence of the heavy ones results in the appearance of an additional plasmon branch. Indeed, from Eqs. (4.16), (4.17), and the condition $S(\mathbf{q}, \omega) = 0$ (poles of Γ) we obtain the dependence $\omega \propto q$, i.e., the new plasmon branch has an acoustic character.

These so-called "demons," acoustic plasmons which are due to the presence of two groups of carriers of different masses, are similar to phonons and can lead to electron pairing. A similar picture can also arise in multivalley semiconductors, semimetals, etc. The possibility of superconducting pairing due to the exchange of demons has been considered in a number of papers ([41], see also the review [42a]).

Exchange of demons as a mechanism of superconductivity in cuprates has been considered in [42b]. Justification of such a mechanism has to be verified experimentally (cf. discussion at the end of Section 4.3).

In this section we discussed collective modes in usual isotropic 3-D metals. The layered conductors display different plasmon modes (see below, Sections 7.2.2.2 and 7.5.1). We think that these modes caused by the presence of low dimensional structural units make an additional (to the phonons) contribution to the pairing in cuprates. We will discuss it in Chapter 7.

4.6.2. *Two-dimensional electron gas*

We consider now plasma oscillations in a two-dimensional electron gas. Such a system can be realized in practice, for example, in an inversion layer where the electronic states are size-quantized in one spatial dimension. This problem has been considered in [43]. We still need to solve Eq. (4.14), but now the Coulomb term has a different form: $V(q) = 2\pi e^2/q$.

In the limit of $\alpha \gg 1$, the polarization operator has the form (4.15). As a result, we find

$$\omega \propto q^{1/2}. \tag{4.18}$$

Thus a reduction in the dimensionality of the system leads to radical changes in the plasmon spectrum. Whereas in the three-dimensional case with a single band there is a gap at $q = 0$ (Eq. (4.15a)), there is no gap in two dimensions: $\omega \to 0$ as $q \to 0$.

A few comments are appropriate here about the layered electron gas, which represents a quasi-two-dimensional system. In this model we neglect small interlayer hopping, but include explicitly the Coulomb interaction between carriers from different layers. It turns out that this highly anisotropic three-dimensional system also contains low-frequency plasmon branches ($\omega \propto q$).

The appearance of superconductivity caused by such plasmons has been studied in [44, 45]. This case is directly related to the high T_c cuprates, and we will discuss it in detail in Chapter 7.

It is interesting that reduced dimensionality leads to the appearance of soft plasmon modes even in a single band picture. A small value of the Fermi energy E_F means the realistic appearance of a narrow band. A charge transfer mechanism based on the coexistence of wide and narrow bands was proposed in [46].

4.7. Coexistence of phonon and electronic mechanisms

We discussed above various electronic mechanisms of superconductivity. In principle, these mechanisms can provide a transition to the super-conducting state without any phonon participation. But it is also perfectly realistic to have a scenario where superconductivity occurs through a joint contribution of the phonon and electronic mechanisms.

Let us analyze this question in more detail. For concreteness, we consider a layered structure. At $T = T_c$, the order parameter describing the pairing satisfies the following equation:

$$\Delta(\kappa, p_z; \omega_n)Z = T \sum_{\omega'_n} \int d\kappa' \, dp'_z \, \Gamma(\mathbf{q}; p_z, p_{z'}; \omega_n - \omega_{n'}) \frac{\Delta(\kappa', p_{z'}; \omega_{n'})}{\omega_n'^2 + \xi'^2}\bigg|_{T_c}, \quad (4.19)$$

where ξ' is the electron energy referred to the Fermi level, and $\mathbf{q} = \kappa - \kappa'$ is the two-dimensional momentum transfer. The quantity Γ can be written as a sum: $\Gamma = \Gamma_{\text{ph}} + \Gamma_{\text{el}}$ where $\Gamma_{\text{ph}} = g^2 D$ is the phonon part and Γ_{el} is the electronic part. In the case of an excitonic mechanism Γ_{el} is approximated as $\lambda_{\text{ex}} \Omega_{\text{ex}}^2 [\Omega_{\text{ex}}^2 + (\omega_n - \omega_{n'})^2]^{-1}$, that is, has a form similar to the phonon Green's function. In the case of plasmons, $\Gamma_{\text{pl}} = V(1 - V\Pi)^{-1}$, see Eq. (4.13).

There are two regions of integration in Eq. (4.19). The first one corresponds to $\alpha = \omega/qv_F < 1$, and the second to $\alpha > 1$. For the first region one can use the static approximation, and it gives a repulsive term. For the second, dynamic, region, one can write (in the plasmon-pole approximation) $\Gamma_{\text{pl}} = \lambda_{\text{pl}} \Omega_{\text{pl}}^2 [\Omega_{\text{pl}}^2 + (\omega_n - \omega_{n'})^2]^{-1}$; this term can be combined with V_c. As a result, we obtain the following equation:

$$\Delta(p_z, \omega_n)Z = T \sum_{\omega'_n} \int dp'_z \, d\Omega \bigg\{ \bigg[\lambda_{\text{ph}}(\Omega, p_z, p'_z) \frac{\Omega^2}{\Omega^2 + (\omega_n - \omega'_n)^2} \\ - V_c \theta(|\omega_n|) - \omega_0) \bigg] \\ + \lambda_{\text{pl}}(\Omega, p_z, p'_z) \frac{\Omega^2}{\Omega^2 + (\omega_n - \omega'_n)^2} \bigg\} \frac{\Delta(\omega'_n, p'_z)}{|\omega'_n|} \bigg|_{T=T_c}.$$

$$(4.20)$$

We see that the electronic mechanism provides additional attraction. It is interesting that the usual Coulomb repulsion and the dynamic attraction are both parts of the same total vertex corresponding to different regions in the (ω, κ) plane. For a rough estimate one can use an expression obtained in the weak coupling approximation (setting $\mu = 0$):

$$T_c = T_c^{\text{ph}} \left(\frac{\tilde{\Omega}_{\text{pl}}}{T_c^{\text{ph}}} \right)^\alpha, \qquad \alpha = \frac{\lambda_{\text{pl}}}{\lambda_{\text{pl}} + \lambda_{\text{ph}}}. \quad (4.21)$$

The large scale of the electronic energy makes the additional contribution noticeable even for small λ_{pl}. We can interpret the same result from a different point of view. Both the additional attractive term and the repulsive term correspond to a large energy scale relative to phonons. Employing the usual step-function method, see Eq. (4.6), we can introduce a pseudopotential μ^* which is weakened by the additional attraction. A similar treatment also can be carried out for the strong-coupling case. Therefore, the additional electronic mechanism manifests itself as a function which decreases μ or even makes it negative. This fact can be detected by tunneling spectroscopy (see Chapter 3).

5

MAGNETIC MECHANISMS

5.1. Introduction

The basic idea that elementary excitations arising from magnetic degrees of freedom (magnons) may affect superconducting pairing between electrons predates the discovery of the superconducting cuprates by about two decades. It was considered in papers by Berk and Schrieffer [1] and Doniach and Engelsberg [2], who showed that incipient ferromagnetic order could suppress phonon-induced s-wave pairing. Pairing in ^{3}He associated with nearly ferromagnetic spin fluctuations in the p-wave channel was proposed by Anderson and Brinkman [3]. For the heavy fermion materials, d-wave pairing due to nearly antiferromagnetic spin-fluctuations was proposed by Scalapino, Loh, and Hirsch [4, 6] and Miyake, Schmitt-Rink, and Varma [5].

We note at the outset that pairing interactions based on the repulsive ($U > 0$) single-band Hubbard model lead to d-wave pairing [4]. It is naturally not surprising that after the superconducting cuprates were found to be high-temperature superconductors [7] with parent compounds showing antiferromagnetic order, various extensions of this result were applied to them and have led to a breathtaking advance in understanding the manifestations of magnetic interactions in highly correlated materials. In the following we give a very cursory overview over this huge field and refer the interested reader to a few of the original publications in this field as well existing summaries specifically aimed at spin-fluctuation-induced superconductivity.

For conceptual convenience, it is useful to classify the very large number of theoretical papers on this subject into two categories by their approach: (1) Fermi liquid based models, and (2) non-Fermi liquid models. In this survey chapter we concentrate on a few of the papers in category (1)—models based on Fermi liquid theory. This is dictated by the fact that Fermi liquid theory is a well-developed and mature framework with no need for a separate conceptual development or justification.

The topics we have somewhat arbitrarily chosen (from the huge number of possibilities) to include in our survey characterize to use those most accessible to the physics community familiar with the current state of understanding in condensed matter theory. Nevertheless, we stress that the advances made in these particular areas are considerable and the results are highly original and novel. All of the models use a version of the Hubbard model as a starting Hamiltonian and then use diagrammatic, perturbative or numerical techniques to study the physical consequences such as various instabilities (spin-density wave (SDW), superconductivity, energy gap formation, etc.), one and two-particle Green's functions, and their relationship to experiment.

The models we review are the spin bag concept proposed by Schrieffer, Wen, and Zhang [8], the t–J model due to Emery [9] and Rice and Zhang [10], and the use of slave bosons due to Kotliar, Lee, and Read [11, 12]. We also give an overview on the work on the two-dimensional Hubbard model [13] by Scalapino et al. In addition, we briefly include the idea of spiral distortions of the antiferromagnetic background for low hole doping due to Shraiman and Siggia [14, 15].

Category (2) is included for completeness. It may be argued that the most radical new ideas in condensed matter theory are found in this area. In particular, the resonant valence bond (RVB) concept proposed and elaborated in a series of papers by Anderson [16] and many others [17] as well as the anyon model due to Kalmeyer and Laughlin and coworkers [18, 19] are very exciting new concepts possibly containing the key to the new materials. However, it seems premature to discuss their current state of development as it is still in a state of rapid change, making a summary very subject to be out of date by the time of publication of this book.

We therefore confine ourselves to a mostly qualitative discussion and refer the interested reader to recent work by the main proponents in this field.

5.1.1. *Localized vs. itinerant aspects of the cuprates*

It is a difficult and not well-understood problem how to describe carrier systems in a regime intermediate to the limits of an itinerant model and a fully localized model. The standard band structure approaches, used for the itinerant case, are based on a single-particle picture and differ in how they include correlations between the single-particle states in terms of some mean field approximation. On the other hand, for the localized case such as magnetic insulators, a spin Hamiltonian provides a starting point for the localized electrons and can describe their collective excitations, the spin-waves, and magnetic order, e.g., ferromagnetic, antiferromagnetic or some other more elaborate magnetic structure. The question arises how to go from one description to the other, or more accurately to deal with situations which have aspects of both.

In this chapter we attempt to give—to the best of our ability—a broad overview of the very large number of papers addressing the role of magnetic effects in the cuprates. The motivation for these approaches arises from the experience and insights gained in dealing with strong electron–electron correlation effects in transition metals and transition metal oxides due to pioneering theoretical work by Mott [20], Anderson [21] and Hubbard [22].

The difficulties have to do with the size of the local electron–electron repulsion, which can be estimated from atomic calculations. Typically—for a Cu 3d electron, for example—such energies are 15–20 eV and thus larger than the one-electron conduction bandwidth. We can visualize the difficulties for the traditional band-picture based on the single-particle approach by using a very simple conceptual model—the Hubbard model [22] in one dimension. This considers a one-dimensional chain of N atoms and includes a nearest-neighbor hopping term t (kinetic energy gained by delocalizing over two sites) and an on-site Coulomb repulsion U between two carriers of opposite spin. The corresponding Hamiltonian for the carriers has the form

$$H_{\text{Hubbard}} = -t \sum_{i,\sigma} (c^+_{i+1,\sigma} c_{i\sigma} + \text{h.c.}) + U \sum_i n_{i\uparrow} n_{i\downarrow} \qquad (5.1)$$

where $c_{i,\sigma}$ and its conjugate are annihilation and creation operators for the carriers obeying anticommutation relations, $-|t|$ is the gain in energy by delocalizing the carriers (hopping integral), and U is the Coulomb repulsion between two carriers of opposite spin on the same site. If we choose a particular filling, for example, one carrier on each site—which in the absence of correlation effects ($U \equiv 0$) corresponds to a half-filled band (half-filled since each momentum k up to the Fermi momentum k_F can accommodate two spin states because of the spin degeneracy)—we can visualize the basic problem for the case of $U \gg |t|$. If we place one carrier per site we fill all sites, and assuming carriers on adjacent sites to have opposite spins we obtain an antiferromagnetic ground state. Now, to move a carrier costs energy U to transfer any electron from its site to the next compared to a gain of $-|t|$, so this state is an antiferromagnetic insulator.

A key question, then, is how to describe the system for a carrier concentration around this special, simple case and to predict whether there is a transition as a function of carrier concentration from the insulating (antiferromagnetic state) to a conducting state for a range of values $0 < \eta < \infty$, where $\eta = |t|/U$.

This question remains open for the three dimensional one-band Hubbard model, although the one-dimensional Hubbard model is well understood and a great deal of progress has been achieved for the two-dimensional Hubbard model, both by analytic and extensive numerical methods, by Scalapino [13] and a large group of collaborators.

A central argument given by the proponents of magnetic models for the cuprates is the experimental fact that the starting materials for two of the superconducting compounds $La_{(2-x)}Sr_x CuO_4$ and $YBa_2Cu_3O_{(7-y)}$ ($y < 0.5$) are the antiferromagnetic insulators La_2CuO_4 and $YBa_2Cu_3O_{(7-y)}$ ($y > 0.5$). The antiferromagnetic phase for the undoped La_2CuO_4 material is well described by the 2D Heisenberg Hamiltonian for spin $s = \frac{1}{2}$, with a reduced magnetic moment per Cu-site of $0.4\mu_B$ due to the transverse quantum fluctuations [23]. In addition, detailed studies of the conduction band structure of the same two compounds have established several important features of the actual orbital states involved in the formation of one-electron bands around the Fermi energy. These, by now generally agreed, features are:

1. The conduction band consists of a strongly hybridized mixture of Cu $d_{(x^2-y^2)}$ and O $2p_{x,y}$ orbitals [24–26].
2. The Madelung energy is important in bringing the donor one-electron energy levels (copper 3d orbitals) within 1–2 eV of the acceptor one-electron levels (oxygen p orbitals) [27–29].
3. Apart from the mixing of the CuO orbitals in the plane, there is also sizable bandwidth (about 0.5 eV) for O–O hopping [30, 31].
4. The out-of-plane orbitals such as Cu d_{z^2} and O p_z orbitals of the apical oxygen atoms linking the CuO planes in the perpendicular direction may also play a role and cannot be neglected in a rigorous treatment [32–34].

The consequences of these four points on the low-energy electronic charge and spin excitations are the following.

Because of point (1) the electronic state formed upon hole doping of the antiferromagnetic 2–1–4 compound, for example, by substituting Sr^{2+} for La^{3+}, is a change in the mixture of (d^9L), where d^9 denotes the Cu^{2+} ion hybridized with an oxygen (L for ligand) hole.

Point (2) is a key requirement for the occurrence of the basic (CuO_2) carrier planes with their short (covalent) Cu–O bond distance of 1.9 Å and the formation of hybridized states between Cu^{2+} and O^{2-} of mixed valence character. The large value of the Madelung energy (about 45 eV) arises from the ionic host lattice with large-valence ions (La^{3+}, Y^{3+}, etc.) as stressed by Matheiss [24]. It was considered in great detail by Torrance and Metzger [27], Kondo [28] and Feiner and deLeeuw [29] in their study of the charge transfer nature of these materials and the occurrence of a phase boundary of metal–insulator character.

Point (3) allows for a second channel for the transport of holes in the CuO_2 planes in addition to CuO hopping. Because of the corner-linked nature of the (CuO_2) checkerboard, it facilitates delocalization in the planes by motion on the oxygen sublattice. We note, however, that as long as antiferromagnetic order is present (small doping), the oxygen and copper spins are paired to the Zhang–Rice singlet [10] and only a correlated motion of both is possible as discussed below.

Point (4) is a very important aspect of the Jahn–Teller character of the copper ion in the cuprates, which was a central point in the minds of Bednorz and Mueller [7] as enhancing electron–phonon interactions. Note that it arises in the following manner. Because of the Jahn–Teller character of the Cu^{2+} ion, the orbital symmetry between degenerate $d_{x^2-y^2}$ and d_{z^2} is removed and as a consequence, the apical oxygen is considerably further away (2.4 Å) than the planar oxygens. Doping and certain lattice modes such as the octahedral tilt mode rehybridize these two orbitals and bring back some orbital degeneracy [32–34]. The latter may play a role in several key physical phenomena such as electron–phonon coupling, orbital relaxation of NMR T_1 processes, and others.

5.2. Fermi liquid-based theories

5.2.1. *The spin bag model of Schrieffer, Wen, and Zhang* [8]

One of the most carefully presented models for the origin of the pairing interaction between the carriers based on magnetic effects is the spin bag model [8]. Physically, the model is presented as a version of the magnetic polaron, in which one carrier locally perturbs the antiferromagnetic order by its own spin, forming a spin bag. A second carrier within the coherence length of this local distortion (estimated to be of order of two or three lattice constants) experiences an attractive interaction by sharing the bag.

In this model it is assumed that BCS pairing theory can be used and that the physical quasi-particles are spin-$\frac{1}{2}$ fermions dressed by a local antiferromagnetic distortion, which is treated in mean-field approximation. The single band 2D Hubbard model is chosen to describe the carriers.

$$H = \sum_{k,\alpha} \varepsilon_k c^+_{k,\alpha} c_{k,\alpha} + \frac{U}{2} \frac{1}{N} \sum_{k,k',q} \sum_{\substack{\alpha,\alpha' \\ \beta,\beta'}} \delta_{\alpha\alpha'} \delta_{\beta\beta'} c^+_{k'\alpha'} c^+_{-k'+q\beta'} c_{-k+q\beta} c_{k\alpha},$$

$$(5.2)$$

$$\varepsilon_k = -2t(\cos \mathbf{k}_x a + \cos \mathbf{k}_y a).$$

Furthermore, it is assumed that one may start from the itinerant (weak coupling) limit of this model, in which $U \ll t$. The half-filled band case (see Fig. 5.1 for its Fermi surface) is known to undergo a spin-density wave (SDW) distortion, which introduces a SDW gap Δ_{SDW} and an antiferromagnetic ground state. For the true ground state $|\Omega\rangle$ with static spin-density wave included, assuming the SDW mean field to be polarized in the z-direction,

$$\langle \Omega | S_Q^2 | \Omega \rangle = \sum_k \langle \Omega | c^+_{k+Q\alpha} \sigma^3_{\alpha\alpha'} c_{k\alpha'} | \Omega \rangle = SN, \qquad (5.3)$$

S is a variational parameter.

The Hartree–Fock form of the starting Hamiltonian (5.2)

$$H_{HF} = \sum_{k,\alpha} \varepsilon_k c^+_{k\alpha} c_{k\alpha} - \frac{US}{2} N \sum_{k,\alpha,\alpha'} c^+_{k+Q\alpha} \sigma^3_{\alpha\alpha'} c_{k\alpha'} \qquad (5.4)$$

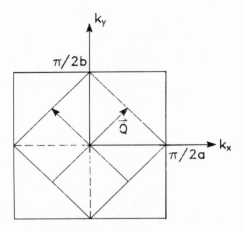

FIG. 5.1. Fermi surface at half-filling; $\mathbf{Q}$ = nesting vector.

is diagonalized by the following transformation to H_{SDW}:

$$H_{SDW} = \sum_{k,\alpha'} E_k(\gamma_{k\alpha}^{+c}\gamma_{k\alpha}^{c} - \gamma_{k\alpha}^{+v}\gamma_{k\alpha}^{v}),$$

$$\gamma_{k\alpha}^{c} = u_k c_{k\alpha} + v_k \sum_{\beta} (\sigma^3)_{\alpha\beta} c_{k+Q\beta},$$

(5.5)

$$\gamma_{k\alpha}^{v} = v_k c_{k\alpha} - u_k \sum_{\beta} (\sigma^3)_{\alpha\beta} c_{k+Q\beta};$$

$$\left.\begin{matrix} u_k \\ v_k \end{matrix}\right\} = \frac{1}{\sqrt{2}}\left(1 \pm \frac{\varepsilon_k}{E_k}\right)^{1/2}, \qquad E_k = (\varepsilon_k^2 + \Delta_{SDW}^2)^{1/2}, \qquad \Delta_{SDW} = \frac{U}{2}S.$$

Δ_{SDW} is the SDW energy gap

These antiferromagnetic correlations at momentum $Q = (\pi/a, \pi/b, 0)$ are built into a set of quasi-particle states, which have a mean-field character and consist of superposition of momentum states k and $k + Q$ are either spin-up (even sublattice) or spin-down (odd sublattice). The next step in the spin bag formalism is the use of the random-phase approximation for the calculation of the various susceptibilities in all available charge and spin channels (Figs. 5.2 and 5.3).

The charge and spin correlation functions (susceptibilities) are defined as

$$\chi^{00}(q, t) = \frac{i}{2N} \langle 0| T[\rho_q(t)\rho_{-q}(0)]|0\rangle,$$

(5.6)

$$\chi^{ij}(q, t) = \frac{i}{2N} \langle 0| T[S_q^i(t)S_{-q}^j(0)]|0\rangle;$$

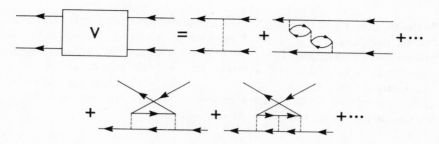

FIG. 5.2.

$$\rho_q = \sum_{k,\alpha} c^+_{k+q\alpha} c_{k\alpha} \qquad \text{charge density operator,}$$

$$S^i_k = \sum_{k,\alpha,\beta} c^+_{k+q\alpha} \sigma^i_{\alpha\beta} c_{k\beta} \qquad \text{spin density operator,} \qquad (5.7)$$

$$\sigma^i_{\alpha\beta} \qquad \text{Pauli matrix.}$$

The next step is the construction of the pair potential for two electrons in the charge and three spin-channels. The spin channels correspond to either

(a)

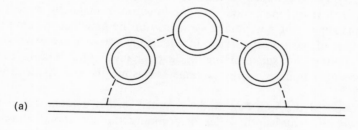

(b)

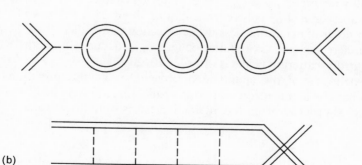

FIG. 5.3. (a) One-loop paramagnon corrections to electron self-energy in SDW state. (b) Feynmann diagrams used in RPA to calculate pairing in SDW background.

fluctuations in the spin-density wave amplitude associated with the longitudinal component (σ_z) or the transverse components corresponding to orientational fluctuations ($\sigma_{+,-}$). It is found that for the parameter range considered the longitudinal fluctuations dominate the pair potential.

Using the random-phase approximation (RPA) the correlation functions in the presence of the Coulomb repulsion U may be expressed in terms of the noninteracting correlation functions $\chi_0^{00}(q, \omega)$,

$$\chi_{RPA}^{00}(\mathbf{q}, \omega) = \frac{\chi_0^{00}(q, \omega)}{1 + U\chi_0^{00}(q, \omega)} \tag{5.8}$$

in the charge channel, and

$$\chi_{RPA}^{ii}(\mathbf{q}, \omega) = \frac{\chi_0^{00}(q, \omega)}{1 - U\chi_0^{00}(q, \omega)} \delta^{ij} \tag{5.9}$$

in the spin channel.

The authors then consider the orbital symmetry of the superconducting order parameter Δ for several forms of the carrier Fermi surface. They point out that it is possible to construct an order parameter with no zeros at the Fermi surface, as seems to be implied by experiments sensitive to such features [35]. This ingeniously overcomes the apparent difficulty of obtaining a p-wave or d-wave superconducting gap-function in magnetic models and experiments suggesting s-wave-like symmetry of the gap. The idea is to use those regions of the Fermi surface, in which the spin-density wave gap is finite for letting the superconducting gap change sign, leading to the appearance of a nodeless order parameter despite its underlying d-wave structure.

As proposed the spin bag model requires certain inequalities to hold for the two different length scales determining the superconducting and antiferromagnetic order. These length scales are set by the superconducting coherence length ξ_{SC} and the antiferromagnetic correlation length ξ_{SDW}. The inequality requires $\xi_{SDW} < \xi_{SC}$. The cuprates are characterized by an extremely short superconducting coherence length of about 25 Å in the CuO_2 planes and an order of magnitude less in the direction perpendicular to the planes. The antiferromagnetic correlation length ξ_{SDW} deduced from the width of the neutron scattering peaks is of order 10–20 Å.

Extensions of the spin bag approach to the interpretation of the photoemission and neutron-scattering results [36, 37] were made in [38, 39] and higher-order terms in the interaction terms were investigated in [40].

5.2.2. The t–J model [9, 10]

Although the initial theoretical papers used the Hubbard model to describe the carrier dynamics in the cuprates with the central role being played by the Cu(d^9, d^8), corresponding to Cu^{2+} and Cu^{3+}, respectively, it was soon discovered from X-ray absorption and electron loss experiments [41–44] that the carriers introduced through doping mainly resided on the oxygens.

This makes necessary the inclusion of the oxygen orbitals in the starting Hamiltonian [9] and increases the complexity of the parameter space. The work of Zhang and Rice [10] was intended to resolve the controversy of whether a single-band or multiband Hubbard model was necessary to describe the essential physics. The starting Hamiltonian used is

$$H = \sum_{i,\sigma} \varepsilon_d d_{i\sigma}^+ d_{i\sigma} + \sum_{l,\sigma} \varepsilon_p p_{l\sigma}^+ p_{l\sigma} + U \sum_1 d_{i\uparrow}^+ d_{i\uparrow} d_{i\downarrow}^+ d_{i\downarrow}, \qquad (5.10)$$

where $d_{i\sigma}^+$ describes Cu $d_{x^2-y^2}$ holes at site i with spin σ, $p_{l\sigma}^+$ creates O($2p_{x,y}$) holes with spin σ at site l. U is the Coulomb repulsion at the Cu site. The orbital energy of the Cu holes is set $\varepsilon_d = 0$ to define the energy scale and the case $\varepsilon_p > 0$ studied. The hybridization term H' is

$$H' = \sum_{i,\sigma l \leftarrow \{i\}} \sum (V_i d_{i\sigma}^+ p_{l\sigma} + \text{h.c.}), \qquad (5.11)$$

where the sum over l is over the four oxygen neighbors of the Cu site at i. The hybridization matrix element V_{il} is proportional to the wave function overlap of the Cu and O holes. Taking account of the phases of the wave functions, one finds

$$V_{il} = (-1)^{M_{il}} t_0, \qquad (5.12)$$

where t_0 is the amplitude of the hybridization and $M_{il} = 2$ if $l = i - 1/2x$ or $i - 1/2y$, $M_{il} = 1$ if $l = i + 1/2x$ or $l = i + 1/2y$. The Cu–Cu distance is used as the length unit. It is further assumed that the following inequalities hold: $t_0 \ll U, \varepsilon_p, U - \varepsilon_p$. Undoped La$_2CuO_4$ has one hole per Cu. At $t_0 = 0$ (atomic limit) and $(U, \varepsilon_p) > 0$ all Cu sites are singly occupied, and all O sites are empty in the hole representation. If t_0 is finite but small, virtual hopping involving doubly occupied Cu-hole states produces a superexchange antiferromagnetic interaction between neighboring Cu holes. The Hamiltonian reduces to an $S = \frac{1}{2}$ Heisenberg model on a square lattice of Cu sites:

$$H_s = J \sum_{ij} (\mathbf{S}_i \cdot \mathbf{S}_j), \qquad J = \frac{4t_0^4}{\varepsilon_p^2 U} + \frac{4t_0^4}{2\varepsilon_p^3}, \qquad (5.13)$$

where S_i are spin-$\frac{1}{2}$ operators of the Cu holes and $\langle ij \rangle$ are nearest neighbor pairs.

Doping introduces additional holes into the CuO$_2$ layers. In the atomic limit $t_0 \to 0$, the additional holes sit at Cu sites if $\varepsilon_p > U$ or at O sites in the opposite case. In the former case, the hybridization may be included by eliminating the O sites to give an effective Hamiltonian for motion on Cu sites only. This is obviously then a single-band Hubbard model. In the second case, it is not so clear that one can eliminate the O sites. We consider this case in the following and show that it also leads to an single-band effective Hamiltonian for the CuO_2 layer.

To explore the second case, consider the energy of an extra hole in La_2CuO_4. To zeroth order in t_0 this energy is ε_p for any O hole state. However, the system may gain energy from Cu–O hybridization, which leads to an antiferromagnetic (AF) superexchange interaction between O and Cu holes. Therefore, the first task is to choose a proper set of the localized O-hole states.

Consider the combinations of four oxygen hole states around a copper ion. They can form either symmetric or antisymmetric states with respect to the central copper ion:

$$P_{i,\sigma}^{(S,A)} = \tfrac{1}{2}\sum_{l\in\{i\}}(\pm 1)^{M_{il}}p_{l\sigma}, \qquad (5.14)$$

where $(\pm)$ correspond to the symmetric S or anti-symmetric (A) state and the phases of the p and d wave functions are defined in Fig. 5.4. Both the S and A states may combine with the d-wave hole to form either singlet or triplet spin states. To second-order perturbation theory, the energies of the singlet and triplet states for S are $-8(t_1 + t_2)$ and 0, respectively, where $t_1 = t_0^2/\varepsilon_p$ and $t_2 = t_0^2/(U - \varepsilon_p)$, while A has energy $-4t_1$. In band structure language, S forms bonding and anti-bonding states, while A is non-bonding (the self-energy of the central Cu^{2+} is $-4t_1$ in the absence of the O hole; the A state has the same energy, so it is nonbonding).

The large binding energy in the singlet S state is due to the phase coherence. This energy should be compared with the energy of an O hole at a fixed site l. In the latter case, the binding energy of a singlet combination

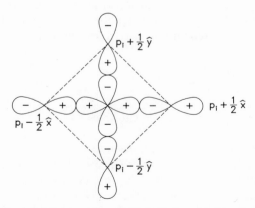

FIG. 5.4. Zhang–Rice hybridization of oxygen hole $(2p^5)$ and copper hole (d^9).

of a O hole and its neighboring Cu hole is only $-2(t_1 + t_2)$, 1/4 of the phase-coherent square S-state. Because the effective hopping energy of the O hole is t_1 or t_2 (depending on the spin configuration), much less than the energy separation of these localized states, we may safely project out the antisymmetric O-hole state, and work in the subspace of the S states. The energy of two O holes residing on the same square, i.e., the configuration $P_i^{(S)} P_i^{(S)} d_{i\sigma}$ is $-(6t_1 + 4t_2)$ much higher than the energy of two separated O holes. It follows that the two holes feel a strong repulsion on the same square.

The localized states of the square are not orthogonal, however, because neighboring squares share a common O site. Therefore,

$$\langle P_{i\sigma}^{(S)} | P_{i\sigma'}^{(S)+} \rangle = \delta_{\sigma\sigma'}(\delta_{ij} - \tfrac{1}{4}\delta_{\langle ij \rangle, 0}). \tag{5.15}$$

In analogy to Anderson's treatment for the isolated spin quasi-particle, one can construct a set of Wannier functions (N_s is the number of squares):

$$\Phi_{i\sigma} = N_s^{-1/2} \sum_k P_{k\sigma} \exp\{i\mathbf{k}\cdot\mathbf{R}_i\},$$

$$P_{k\sigma} = N_s^{-1/2}\beta_k \sum_i P_{i\sigma}^{(S)} \exp\{-i\mathbf{k}\cdot\mathbf{R}_i\}, \tag{5.16}$$

and β_k is a normalization factor

$$\beta_k = [1 - \tfrac{1}{2}(\cos k_x + \cos k_y)]^{-1/2}. \tag{5.17}$$

The functions $\Phi_{i\sigma}$ are orthogonal, and complete in the symmetric hole space. $\phi_{i\sigma}$ combines with Cu hole at site i to form a spin singlet $(-)$ or spin triplet $(+)$:

$$\Psi_i^{\pm} = \tfrac{1}{2}(\phi_{i\uparrow} d_{i\downarrow} \pm \phi_{i\downarrow} d_{i\uparrow}), \tag{5.18}$$

with energies in second-order perturbation theory of

$$E_{\pm} = \sum_{\{w\}} |\langle \Psi_i^{\pm}|H|w\rangle|^2/\Delta E_w, \tag{5.19}$$

where w runs over all possible intermediate states, and ΔE_w is the zeroth-order energy difference between Ψ and w, i.e., $\Delta_w = \varepsilon_p - U$ or ε_p depending on whether or not the state w contains a doubly occupied Cu hole. For the sake of simplicity, the parameters are further simplified by choosing $\varepsilon_p = U - \varepsilon_p$; hence $t_1 = t_2 = t$, and the physics is expected to be essentially the same. One finds for the singlet and triplet energies:

$$E_{\pm} = -8(1 \mp \lambda^2)t, \tag{5.20}$$

$$\lambda = N_s^{-1}\Sigma\beta_k^{-1} \approx 0.96. \tag{5.21}$$

These energies are very close to those of a single square. Since $E_+ - E_- = 16t \gg t$, one may safely ignore transitions between $\{\psi_i^-\}$ and $\{\psi_i^+\}$ and the system can be treated within the singlet $\{\psi_i^-\}$ subspace. It is important to stress that it is the phase coherence over the four oxygen sites that produces the large energy separation of the different symmetry states from the spin-singlet state of the Cu hole and the symmetric O hole.

After having obtained the proper Wannier functions with large binding energy, the next problem is the motion of these singlet states due to hopping. Since the O holes are superposed on a background of singly occupied Cu holes, the motion is correlated in the sense that if the state ψ^- moves from site j to i, a Cu hole simultaneously moves from site i to j. This process is represented by $\psi_i^- d_{j\sigma} \to \psi_i^- d_{i\sigma}$ with kinetic energy described by an effective hopping Hamiltonian

$$H_t = \sum_{i \neq j, \sigma} t_{ij} (\Psi_j^- d_{i\sigma})^+ \Psi_j^- d_{j\sigma}, \tag{5.22}$$

with the effective hopping matrix element t_{ij} given within second-order perturbation theory by

$$t_{ij} = \sum_{\langle w \rangle} \langle \Psi_j^- d_{j\sigma} | H' | w \rangle \langle w | H' | (\Psi_j^- d_{i\sigma})^+ \rangle / \Delta E_w. \tag{5.23}$$

We can evaluate t_{ij} in the original O-site representation $p_{l\sigma}$ and find two different types of two-step hopping processes contributing to t_{ij}. One involves spin exchange between the Cu and O holes and is expressed by $t_{ij}^{(a)}$. The second is the effective O-hole hopping and is found for $i \neq j$ as

$$t_{ij} = t_{ij}^{(a)} - \tfrac{1}{2} t \delta_{\langle ij \rangle_0},$$

$$t_{ij}^{(a)} = \frac{8\lambda t}{N_s} \sum_k \beta_k^{-1} \exp \langle i\mathbf{k} \cdot (i - j) \rangle. \tag{5.24}$$

Evaluating these equations for nearest neighbors $i, j, t_{ij} \approx -1.5t$ and all other effective hopping matrix elements are very small, i.e., the next-nearest-neighbor $t_{ij} \simeq -0.16t$, one order of magnitude smaller.

We see that when a Cu d-hole is created at site i, the singlet state is destroyed at the same site. It follows that the state ψ_i^- it is equivalent to the empty state of the d-hole at site i. The effective hopping Hamiltonian, after dropping the empty state operators, is then reduced to the form

$$H_t = \sum_{i \neq j, \sigma} t_{ij} (1 - n_{i-\sigma}) d_{i\sigma}^+ d_{j\sigma} (1 - n_{j,\sigma}). \tag{5.25}$$

As was discussed in the beginning, an effective Heisenberg Hamiltonian $H_J = J \sum \mathbf{S}_i \cdot \mathbf{S}_j$ holds for the doped system describing the antiferromagnetic interaction between the d-holes. The singlet state has no magnetic interaction with all other d-holes. In summary, an effective Hamiltonian H_{eff} has been found,

$$H_{eff} = H_t + H_J, \tag{5.26}$$

where both H_t and H_J refer only to Cu d-holes. This once again is the effective Hamiltonian of the single-band Hubbard model in the large-U limit.

5.2.3. Two-dimensional Hubbard model studies by Monte Carlo techniques

A different approach to the study of the role of magnetic excitations in the layered cuprates has been pioneered by Scalapino and a large number of collaborators [13, 45]. It is an attractive alternative to the field-theoretic and phenomenological approaches discussed so far as it uses numerical methods such as the quantum Monte Carlo simulation of the partition function and various correlation functions for a given model Hamiltonian such as the Hubbard model on a finite square lattice. The results obtained are exact for the size of lattice considered (currently 10×10 is typical maximum size) and can offer considerable insight into the short-range correlations induced in the appropriate correlation functions. Even with the relatively small size lattices these calculations are numerically very intensive as the number of configurations sampled varies as $N^3 L$.

It is assumed at the outset that a two-dimensional single-band Hubbard model describes the basic physics of the strongly correlated carriers in both the undoped (antiferromagnetic) starting materials (La_2CuO_4, $YBa_2Cu_3O_6$) and the doped, conducting and superconducting materials ($La_{(2-x)}Sr_xCuO_4$, $YBa_2Cu_3O_{(7-y)}$, etc.).

The Hubbard model is simulated on a finite $N \times N$ lattice by calculating the partition function Z, the one-particle and two-particle Green functions, and other quantities of physical interest which can be related to them. A very clear description of this approach was given by Scalapino at the Santa Fe meeting [45] on correlated systems, which we follow here.

It is instructive to point out that even a simple-looking model such as the single-band Hubbard model (5.1) which depends only the parameters U/t, $n_i = \langle n_{i\uparrow} + n_{i\downarrow} \rangle$, the band filling, and the temperature T is amazingly rich in content, and as we will see is successful in instructing us about some of the physics occurring in strongly correlated systems.

The key quantity to be calculated is the partition function Z defined as

$$Z = \text{Tr}\, e^{-\beta(H-\mu N)} = \text{Tr}\, e^{-\Delta\tau H} e^{-\Delta\tau H} \cdots e^{-\Delta\tau H}, \tag{5.27}$$

where we have divided the interval $(0, \beta)$ into N imaginary time slices $\Delta\tau$ such that $N\,\Delta\tau = \beta = 1/k_B T$ and T is the temperature. Assuming $\Delta\tau$ is small, we have approximately

$$e^{-\Delta\tau H} \approx e^{-\Delta\tau K} e^{-\Delta\tau V} + O(\Delta\tau^2 tU), \tag{5.28}$$

where K is the first and V the second term in Eq. (5.1). Using the ideas introduced by Stratonovich [46] and Hubbard [47] (Wang, Evanson, and Schrieffer [48]), it is possible to define a Hubbard–Stratonovich field $x_i(\tau_l)$ at site i, so that the bilinear term in n_i proportional to U may be written

$$\exp(-\Delta\tau U n_{i\uparrow} n_{i\downarrow}) = \left(\frac{\Delta\tau}{\pi}\right)^{1/2} \exp[(\Delta\tau U/2)(n_{i\uparrow} + n_{i\downarrow})] \int_{-\infty}^{\infty} dx_i(\tau_l)$$

$$\times \exp\{-\Delta\tau[x_i^2(\tau_l) + (2U)^{1/2} x_i(\tau_l)(n_{i\uparrow} - n_{i\downarrow})\}. \qquad (5.29)$$

The partition function Z can then be expressed as

$$Z = \sum_{\langle x_i(\tau_l) \rangle} \text{Tr } e^{-\Delta\tau h(x(\tau_n))} \cdots e^{-\Delta\tau h(x(\tau_1))}, \qquad (5.30)$$

where $h(x(\tau_l)) = \sum c_i^+ h_{ij}(x(\tau_l)) c_j$ is an effective one-body interaction, which has to be averaged over all Hubbard–Stratonovich fields $h_{ij}(x(\tau_l))$. The remarkable advantage of his transformation is the reduction of the original Hamiltonian containing two-body interactions to an effective one-body Hamiltonian. The latter may, of course, be solved exactly, although we note that we have paid a certain price, namely the presence of the effective Hubbard–Stratonovich field $h_{ij}(x(\tau_l))$, which has to be averaged over all values.

Carrying out the fermion trace in Eq. (5.30) one finds

$$Z = \sum_S \det M_\uparrow(\langle S \rangle) \det M_\downarrow(\langle S \rangle), \qquad (5.31)$$

where $M_\sigma(S) = (1 + e^{\beta\mu} \prod_1 e^{-\Delta\tau h(1,\sigma)})$. The Hubbard–Stratonovich reformulation of the original problem can be used as a starting point for stochastic sampling procedures. The product of the spin-up and spin-down Fermion determinants acts as a weight function for Monte Carlo importance sampling just as the Boltzmann weight would do in a classical calculation. Note that it is a key requirement for this approach to be valid that the sign of the weight function $P(\{S\})$ is positive definite. We will return to this aspect later in a discussion of the problem areas of the Monte Carlo approach. The weight function $P(\{S\})$ is defined as

$$P(\{S\}) = \frac{\det M_\uparrow(\langle S \rangle) \det M_\downarrow(\langle S \rangle)}{\sum_{\langle S' \rangle} \det M_\uparrow(\langle S' \rangle) \det M_\downarrow(\langle S' \rangle)} \qquad (5.32)$$

and generates configurations in a Monte Carlo algorithm distributed according to this weight. It is then possible to calculate with these configurations thermodynamic one- and two-particle Green's functions and equal-time expectation values.

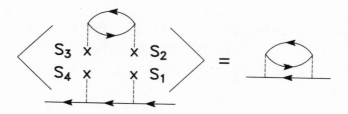

FIG. 5.5.

To demonstrate the power and usefulness of this method consider the one-particle Green's function

$$G_{ij}(\tau) = -\text{Tr}\, e^{-\beta H} T\langle c_{i\sigma}(\tau) c_{j\sigma}^{+}(0)\rangle/Z. \qquad (5.33)$$

It can be found by summing the different expressions

$$G_{ij}(\tau; \langle S\rangle) = (1 + e^{\beta\mu}\, e^{-\Delta\tau h(l-1,0)} \ldots e^{-\Delta\tau h(1,0)})^{-1} \qquad (5.34)$$

over configurations $\{S\}$ with weight factor $P(\{S\})$. Then define the Monte Carlo average of G_{ij} below:

$$\bar{G}_{ij} = G_{ij}(\tau_j\langle S\rangle)_{MC} = \frac{1}{M}\sum_S G_{ij}(\tau_j\langle S\rangle), \qquad (5.35)$$

where M is the number of Monte Carlo configurations. Figure 5.5 shows schematically how this sum, in principle, includes all Feynman diagrams. The simulation is, therefore, nonperturbative and the results are exact except for errors (of order $\Delta\tau^2$) due to the finite step-size $\Delta\tau$. It is, of course, essential that all regions of phase space are sampled.

Other quantities of physical interest are the spatial Fourier transform of $G_{ij}(\tau)$, denoted $G_p(\tau)$, and its limit $\tau \to 0-$, which is the mean-particle number $\langle n_p\rangle$.

The kinetic energy in the grand canonical ensemble average is given by

$$\langle T\rangle = \sum_{p,\sigma} \varepsilon_p G_\sigma(p, \tau \to 0). \qquad (5.36)$$

One can also evaluate the Fourier transform of $G_p(p, \tau)$ with respect to τ at the Matsubara frequencies $\omega_n = (2n + 1)\pi T$,

$$G(p, i\omega_n) = \int_0^\beta d\tau\, e^{i\omega_n\tau}\, G(p, \tau) = \frac{1}{i\omega_n - (\varepsilon_p - \mu) - \Sigma(p, i\omega_n)}, \qquad (5.37)$$

and exact information on the exact self-energy $\Sigma(p, i\omega_n)$.

Consider the evaluation of a two-particle Green's function within the Monte Carlo simulation approach. Specifically, define the s-wave pair field propagator $D(\tau)$ needed to describe superconducting correlations:

$$D(\tau) = -\langle T[\Delta(\tau)\Delta^+]\rangle, \qquad (5.38)$$

where the operator Δ is defined by

$$\Delta = \left(\frac{1}{N^{1/2}}\right)\sum_l c_{i\uparrow}c_{i\downarrow}. \qquad (5.39)$$

For a given Hubbard–Stratonovich configuration one calculates $D(\tau)$ by letting the up- and down-spin operator interact with the effective magnetic field given by the Hubbard–Stratonovich distribution. One finds

$$\frac{1}{N}\sum_{ij}\int_0^\beta G_{ij\uparrow}((\tau); \langle S\rangle)G_{ij\downarrow}(\tau; \langle S\rangle)\,d\tau. \qquad (5.40)$$

Next this has to be averaged over a set of Monte Carlo configurations $\{S\}$ with their appropriate weight $P(\{S\})$ to give the exact result (for a graphical representation see Fig. 5.6)

$$D = \frac{1}{N}\sum_{ij}\int_0^\beta \langle G_{ij\uparrow}(\tau_i\langle S\rangle)G_{ij\downarrow}(\tau_i\langle S\rangle)\rangle_{MC}\,d\tau. \qquad (5.41)$$

Note that the first part of Fig. 5.6 corresponds to the product of the average of two one-particle Green's functions, while the second part includes all effects of the pairing interaction. The two figures and Eqs. (5.36–5.41) demonstrate a direct correspondence between the method of thermodynamic Green's functions and the Monte Carlo simulation, which allows the numerical study of these quantities in an exact form.

Having described the advantages and very attractive features of this approach, it seems fair to also discuss its drawbacks and limitations. The first is the constraint to finite and currently quite small (10×10 is a typical value) spatial lattices. The answers found need to be extrapolated to the bulk limit and, as the problem under study is a quantum mechanical system, to zero temperatures ($\beta = L\delta\tau \to \infty$). In addition to the problems, which are similar to those arising in simulating classical systems, there are distinctly new difficulties for the quantum case, making it numerically much more extensive.

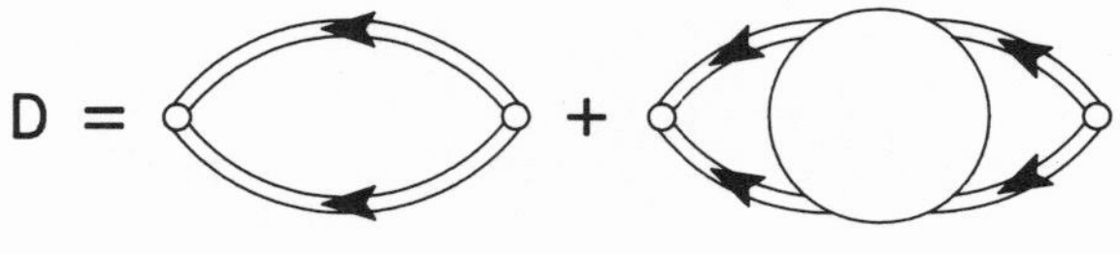

FIG. 5.6.

If we compare a classical problem with short-range interactions, the number of arithmetic steps for a Monte Carlo update at each site is of order N, the number of sites. For the Hubbard model, on the other hand, the nonlocal nature of the fermion determinant leads to $(NL)^3$ operations per site, hence a total of $(NL)^4$ operations for a space–time sweep. It has been possible by a refinement of algorithms to reduce this to N^3L operations. Still, the residual increase of a factor N^2 for a typical $N \times N$ lattice contributes a factor of 10^4 to the computation for current lattice sizes.

In addition, the fermion matrices B,

$$B = \prod_l \exp[-\Delta\tau h(l, \sigma)], \qquad (5.42)$$

which are needed in the calculation become ill-conditioned as the temperature is lowered, as the eigenvalues of B span a huge range between $\exp(-40)$ and $\exp(40)$ for a temperature $T = 0.1t$. Again techniques have been developed to separate different energy scales and avoid swamping the small eigenvalues by round-off errors.

Finally, the most daunting difficulty in fermion simulations is the "sign" problem. It arises from the fact that although the partition function is positive, there is no requirement for the contributions from specific configurations also to be so. As a consequence the product $\det M_\uparrow(\{S\}) \det M_\downarrow(\{S\})$ can fluctuate in sign. Only the presence of particle–hole symmetry for the half-filled Hubbard model, for example, guarantees the product of the two determinants for up and down spins to be positive definite.

If the product $\det M_\uparrow \det M_\downarrow$ changes sign, the probability distribution is taken proportional to the absolute value of the product,

$$\tilde{P}(\{S\}) = \frac{|\det M_\uparrow(\langle S \rangle) \det M_\downarrow(\langle S \rangle)|}{\sum^{S'} |\det M_\uparrow(\langle S' \rangle) \det M_\downarrow(\langle S' \rangle)|}, \qquad (5.43)$$

and expectation values of operators O are calculated as

$$\langle O \rangle = \frac{\langle O \, \text{sgn}(\tilde{P}) \rangle_{\tilde{P}}}{\langle \text{sgn}(\tilde{P}) \rangle_{\tilde{P}}}. \qquad (5.44)$$

Here the subscript $\tilde{P}$ indicates an average taken with the distribution function of Eq. (5.43). Note that if $\langle \text{sgn} P \rangle$, the average sign, becomes small, the Monte Carlo results will show large fluctuations. This is particularly troublesome for low temperature as β becomes large and the average sign decays exponentially with increasing β. This was also found, in general, for a variety of band fillings and interaction strengths U studied. It is most unfortunate that just below half-filling the average sign decreases rapidly and makes the study of this physically very relevant region in the cuprates very difficult.

Finally, there is a difficulty in numerical simulations in extracting real frequency-domain results from the imaginary time data. Although the formal route of simply analytically continuing from $i\omega_n \to \omega + i\delta$ seems easy, the analytical continuation of numerical data is an ill-conditioned problem extremely sensitive to statistical fluctuations.

In the following, some examples of calculated quantities are given and compared with results obtained from other studies. The local moment is defined $m_{zi} = n_{i\uparrow} - n_{i\downarrow}$. Its average square is given by

$$\langle m_{zi}^2 \rangle = \langle n_{i\uparrow} + n_{i\downarrow} \rangle - 2\langle n_{i\uparrow} n_{i\downarrow} \rangle. \tag{5.45}$$

At half-filling $\langle n_{i\uparrow} \cdot n_{i\downarrow} \rangle = 1$ and the second term in (5.45), which measures double occupation, will decrease as U increases. For example, when $U = 8t$ (the band width for two dimensions) the probability of double occupancy is 20% less than for $U = 0$.

Next consider the effective one-electron transfer term

$$\frac{t_{\text{eff}}}{t} = \frac{\langle c_{i\sigma}^+ c_{j\sigma} + c_{j\sigma}^+ c_{i\sigma} \rangle_U}{\langle c_{i\sigma}^+ c_{j\sigma} + c_{j\sigma}^+ c_{i\sigma} \rangle_0}. \tag{5.46}$$

For weak coupling, one finds analytically

$$\frac{t_{\text{eff}}}{t} = 1 - \left(\frac{U}{t}\right)^2 \frac{\Delta\varepsilon_2}{\Delta\varepsilon_0}, \tag{5.47}$$

where $\Delta\varepsilon_0$ and $\Delta\varepsilon_2$ are functions of the band energy and Fermi functions. For strong coupling, on the other hand, one has

$$\frac{t_{\text{eff}}}{t} = \frac{4(4t/U)(|\langle \mathbf{S}_i \cdot \mathbf{S}_j \rangle| + \frac{1}{4})}{\Delta\varepsilon_0}. \tag{5.48}$$

For the effective hopping term one can calculate the integral of the optical weight $\sigma(\omega)$,

$$8\int_0^\infty d\omega \, \sigma(\omega) = 4\pi e^2 E_0(t_{\text{eff}}/t) \tag{5.49}$$

where E_0 is the energy per site for $U = 0$ at a given band filling $\langle n \rangle$. One notes that at half-filling this will go to zero with large U. For $\langle n \rangle = 1$, the integrated spectral weight increases as the vacancies created decrease the effective mass of the carriers. As an example of long-range effects, consider the equal-time magnetic moment correlation function

$$C(l) = \langle m_{i+l}^z m_i^z \rangle. \tag{5.50}$$

One finds that for the half-filled case $\langle n \rangle = 1$ the antiferromagnetic correlations extend throughout the lattice, while for quarter-filling the correlations decay within one lattice spacing.

From the knowledge of $C(l)$, one can calculate the magnetic structure factor $S(q)$:

$$S(q) = \sum_l \exp(iq \cdot l) \langle m_{i+l}^z m_i^z \rangle \tag{5.51}$$

For the half-filled case, the peak at the antiferromagnetic wave vector (π, π) is very visible. For the case of long-range antiferromagnetic order,

$$\lim_{N \to \infty} S(\pi, \pi)/N = m^2/3, \tag{5.52}$$

where m is the antiferromagnetic order parameter. One can obtain m by extrapolating the spin–spin correlation function to infinite lattice spacing. For $U > 8t$, m is found to reach the value found for the $S = \frac{1}{2}$ Heisenberg antiferromagnet [49].

For doping off half-filling, the sign problem discussed earlier presents difficulties. However, antiferromagnetic correlations are rapidly suppressed away from half-filling. The antiferromagnetic peak in the structure factor becomes smaller and seems to split into two peaks at $(\pi, \pi - \Delta_q)$ and $(\pi - \Delta_q, \pi)$. For the accessible temperature range $\beta < 6$, there is no evidence of growing incommensurate order.

Here we discuss the information content of the single-particle Green's function $G_{ij}^\sigma(\tau)$ which is related to several single particle properties:

$$G_{ij}^\sigma(\tau) = \langle T c_{i\sigma}(\tau) c_{j\sigma}^+(0) \rangle. \tag{5.53}$$

The physical quantities one may calculate are the single-particle momentum distribution $\langle n_{k\sigma} \rangle = \sum_i \exp\{ik \cdot l\} G_{i+l,i}^\sigma(0^-)$, the band filling $\langle n \rangle$,

$$\langle n \rangle = \frac{1}{N} \Sigma G_{l,i}^\sigma(0^-), \tag{5.54}$$

the chemical potential μ, and the compressibility $K = (1/n^2) \, dn/d\mu$.

Another very important quantity is the one-electron self-energy $\Sigma(k, i\omega_n)$, which can be calculated from the Fourier transform of G at the Matsubara frequencies ω_n:

$$\Sigma(k, i\omega_n) = i\omega_n - (\varepsilon_k - \mu) - G^{-1}(k, i\omega_n). \tag{5.55}$$

Experience with calculations has shown that $G_{ij}^\sigma(\tau)$ has the smallest statistical fluctuations in Monte Carlo simulations. It is, therefore, very helpful for obtaining one-electron results on fairly large lattices.

Several examples are given for $\langle n_k \rangle$ for different sized lattices at $k = (\pi/2, \pi/2)$ $U = 4$ and $\beta = 6$; $6 \times 6 \to 14 \times 14$. The noninteracting Fermi function $f(\varepsilon_k) = (\exp\{-(\varepsilon_k - \mu)\} + 1)^{-1}$ has been used with $\varepsilon_k = -2t(\cos k_x + \cos k_y)$, while the chemical potential μ has been adjusted to give $\langle n \rangle = 1, 0.5$ for the half- and quarter-filled band, respectively. We also show the mean field result

$$\Delta = \frac{1}{N} \sum_k \frac{1}{E_k} \qquad (5.56)$$

It is found that for the 16×16 lattice the mean-field gap is too large and one needs a reduced gap near $k = (\pi/2, \pi/2)$.

The compressibility K is a good measure of the metallic or insulating character of the 2D Hubbard model. For $U = 0$,

$$K = \frac{2}{NT} \sum_k f(\varepsilon_k)(1 - f(\varepsilon_k)), \qquad (5.57)$$

and for small T, $K = N(\mu)$, the single-particle density of states:

$$K = \frac{\ln(16t/\mu)}{\pi^2 t}.$$

The divergence of K as $\mu \to 0$ disappears for finite $U = 0$ as a gap opens in the single-particle spectrum.

The behavior of $\Sigma(k, i\omega_n)$ is very interesting. Mean-field theory gives for half-filling

$$\Sigma(k, i\omega_n) = \frac{\Delta^2}{i\omega_n + \varepsilon_k}. \qquad (5.58)$$

For a noninteracting Fermi gas at $\langle n \rangle = 1$, where $\varepsilon_k = 0$, the imaginary part of Σ has the form

$$\text{Im } \Sigma(k, i\omega_n) = -\frac{\Delta^2}{\omega_n}. \qquad (5.59)$$

This is different from the case of a Fermi liquid with

$$\text{Im } \Sigma(k, i\omega_n) = (1 - Z(k, i\omega_n))\omega_n. \qquad (5.60)$$

Here $Z^{-1}(k_F, i\pi T \to 0)$ is the jump in the occupation, which defines the

Fermi surface. As $Z(k_F, i\pi T \to 0) > 1$, for a Fermi liquid Im $\Sigma(k, i\omega_n)$ has a negative slope for small ω_n different from the divergence in the presence of a gap.

From the Monte Carlo simulations, $\Sigma(k, i\omega_n)$ for the half- and quarter-filled cases ($k = (0, \pi)$, $k = (\pi/2, \pi/2)$) has been found and shows qualitatively quite different behavior. For $U = 4$ the quarter-filled system behaves like a Fermi liquid.

One can also extract real frequency information from the Monte Carlo data for

$$N(\omega) = -\tfrac{1}{4} \text{Im } G_{ii}(i\omega_n \to \omega + i\delta) \tag{5.61}$$

and for $U = 4$ and 8 find evidence for a gap. For $U < 2$ numerical difficulties make it impossible to draw unambiguous conclusions. Also the single particle spectral weight shows a gap, which increases with increasing U.

Of very great interest is what can be learned about superconducting correlation functions like the equal-time pair-field correlation function

$$D(l) = \langle \Delta_{l+i,i} \Delta_l^+ \rangle \tag{5.62}$$

and its structure factor

$$D(q) = \sum_l \exp(iq \cdot l) D(l). \tag{5.63}$$

The static pair field susceptibility is

$$P = \int_0^\beta d\tau \langle \Delta(\tau) \Delta^+(0) \rangle. \tag{5.64}$$

The pair-field s-wave operator is $\Delta_l = c_{l\uparrow} c_{l\downarrow}$ and

$$\Delta = \frac{1}{N} \sum_l \Delta_l. \tag{5.65}$$

Also of interest are more general rotational symmetries like extended s-wave (s*) and d-wave fields

$$\begin{aligned}
\Delta_{s*}(l) &= \tfrac{1}{2}(c_{l\uparrow} c_{l+x\downarrow} - c_{l\downarrow} c_{l+x\uparrow}) + \tfrac{1}{2}(c_{c\uparrow} c_{l+y\downarrow} - c_{l\downarrow} c_{l+y\uparrow}), \\
\Delta_d(l) &= \tfrac{1}{2}(c_{l\uparrow} c_{l+x\downarrow} - c_{l\downarrow} c_{l+x\uparrow}) - \tfrac{1}{2}(c_{l\uparrow} c_{l+y\downarrow} - c_{l\downarrow} c_{l+y\uparrow}),
\end{aligned} \tag{5.66}$$

which in momentum space have the form

$$\Delta_{s*} = \left(\frac{1}{N}\right)\frac{1}{2}\sum_l \Delta = \frac{1}{\sqrt{N}} \sum_p (\cos p_x + \cos p_y) c_{p\uparrow} c_{-p\downarrow},$$

$$\Delta_d = \left(\frac{1}{N}\right)\frac{1}{2}\sum (\cos p_x - \cos p_y) c_{p\uparrow} c_{-p\downarrow}. \tag{5.67}$$

5.2.4. *Spiral phase of a doped quantum antiferromagnet* [14, 15]

A low density of vacancies in a 2D, spin-$\frac{1}{2}$ Heisenberg antiferromagnet leads (for a range of couplings) to a metallic phase with *incommensurate* antiferromagnetic order. This order corresponds to the staggered magnetization rotating in the plane with a wavenumber proportional to the density. It arises from the polarization of the antiferromagnetic dipole moments of the vacancies. This spiral state contains an interesting low-lying mode in its excitation spectrum, which has consequences for neutron scattering and the normal state resistivity.

CuO based compounds contain electrons with hard-core interactions on a 2D square lattice at a density of just less than one per site. Residual interactions are antiferromagnetic (AF) and are described by a t–J model Hamiltonian with vacancy hopping and spin-exchange terms:

$$\mathcal{H}_0 = -t \sum_{r,\sigma\hat{a}} c^+_{r+\hat{a},\sigma} c_{r,\sigma} + J \sum_{r,\hat{a}} \mathbf{S}_r \cdot \mathbf{S}_{r+\hat{a}}. \tag{5.68}$$

$\hat{a} = \hat{x}, \hat{y}$, the spin-$\frac{1}{2}$ Fermion operator $c_{r\sigma}$ is restricted to single occupancy, and $s_r = c^+_{r,\sigma} \tau_{\sigma\sigma'} c_{r\sigma}$ is the local spin operator.

In the absence of vacancies, the spin-$\frac{1}{2}$ Heisenberg model has long range AF order at $T = 0$ [23]. It can be described by the nonlinear σ model for its long-wavelength properties. The ground state of a single vacancy is well understood. It forms a narrow band ε_k of width $W \sim O(J)$ if $t \gg J$ (or $W \sim O(t^2/J$ for $t \ll J)$ with energy minima lying at $k_v = (\pm\pi/2, \pm\pi/2)$. The band is very anisotropic with mass $\mu_\perp \sim 1/W$, for k perpendicular to the zone-boundary is about a factor of 10 smaller than the parallel mass, $\mu_\parallel$, for $t = J$. Near the energy minima, the vacancy states involve a long range dipolar distortion of the staggered magnetization, $\hat{\Omega}$, and can be assigned an AF dipole moment $p_\alpha(k)$, which is a vector in both spin and physical space with magnitude of order min$(J^{-1}t, 1)$ and k-dependence $p_\alpha(k) \sim \sin k_\alpha$.

Next consider the effect of a small number of vacancies, $n \ll 1$. The *commensurate* Néel state is unstable (at $T = 0$ in 2D) for any n, towards a spiral state, in which the AF dipole moments of the holes order as a result of the polarization of the spins of the opposite values of the Fermi sea (located near k_v) in opposite directions. In the spiral state, the staggered magnetization, $\hat{\Omega}$, rotates in a plane with a pitch scaling as $1/n$. There is also a *new*, low-lying excitation, the "torsion" mode, which is the collective mode associated with the transverse fluctuations of the dipole polarization, and hence with the fluctuations of the plane of the spiral in spin space. It also could be interpreted in terms of a rotation about the local $\hat{\Omega}$ axis, which appears as an additional degree of freedom for *noncollinear* magnetic structures. We expect the AF correlation length (in some temperature range) to scale like $1/n$. The spiral phase is quite different from a $2k_F$ spin-density wave instability of the Fermi surface, and remains metallic with the vacancy Fermi surface ungapped.

These one-particle properties are included in a semiphenomenological effective Hamiltonian:

$$\mathcal{H}_{\text{eff}} = \sum_k \varepsilon_k \bar{\Psi}_k \Psi_k - g \sum_{q,a} \mathbf{p}_a(\mathbf{q}) \cdot \mathbf{j}_a(\mathbf{q}) - g' \sum_{q,k,a} \cos k_a \bar{\Psi}_{k-q/2} \hat{\tau} \Psi_{k+q/2} \cdot \mathbf{m}(q) + \mathcal{H}_{\text{NL}\sigma}.$$

(5.69)

Here the vacancy is represented by a two-component spinor—the pseudo-spin of the vacancy arises from the sublattice structure induced by local AF correlation of the spin background. The spinor Ψ^α derived from a microscopic decomposition $c_{r,\sigma}^{\alpha+} = \Psi_r^{\alpha+} z_{r,\sigma}^\alpha$, $\Psi_r^{\alpha+}$ creates a Fermionic hole and the spinor $z_{r,\sigma}^\alpha$ is a Schwinger spin boson. The explicit sublattice index $\alpha = $ A, B accommodates the staggered order (and labels the two-fold degeneracy of the vacancy ground state). The hopping part of $\mathcal{H}_0$ has the form

$$\mathcal{H}_{\text{int}} = -t \sum_{\langle r,r' \rangle} (\Psi_r^{+\text{B}} \Psi_{r'}^\text{A} \bar{z}_r^\text{A} z_r^\text{B} + \text{h.c.}).$$

(5.70)

Define $\Psi^\alpha = h_{\alpha\alpha'} \Psi^{\alpha'}$, $\hat{R}$ is an SU(2) rotation relating the spinor z_r and hence the local direction of the staggered magnetization $\hat{\Omega}(r) = \bar{z}_r^\text{A} \tau z_r^\text{A}$ to a fixed basis.

The relation between $\mathcal{H}_{\text{int}}$ and Eq. (5.68) becomes more transparent in the spin-wave approximation after introducing $z_r^\text{A} = (1, a_r) z_r^\text{B} = (b_r, 1)$, transforming to the momentum representation, and identifying the staggered magnetization mode as $(a_q^+ - b_{-q})$ and the net magnetization as $(a_q^+ + b_{-q})$. $\mathbf{m}$ is the local magnetization operator, which is conjugate to $\hat{\Omega}$ and enters in the nonlinear sigma model Hamiltonian,

$$\mathcal{H}_{\text{NL}\sigma} = \tfrac{1}{2} \sum_r (\chi^{-1} m^2 + \rho (\nabla_\alpha \Omega)^2),$$

(5.71)

where $\chi \sim 1/J$ is the susceptibility, $\rho \sim J$ is the spin-wave stiffness, and ∇_α is the lattice gradient. The second term in H_{eff} couples the background magnetization current $j = \hat{\Omega} \times \nabla_\alpha \hat{\Omega}$ with the AF dipole moment of the vacancies,

$$\hat{p}_a(q) = \sum_k \sin k_a \bar{\Psi}_{k-q/2} \hat{\tau} \Psi_{k+q/2},$$

(5.72)

and gives rise to the dipolar interactions.

The phenomenological coupling constants g and g' are of order $\min(t, J)$. Note $g = g' = t$ and $\varepsilon_k = 0$ would correspond to the bare hopping term of H_0; diagrammatically $\varepsilon_k = 0$ emerges as the coherent part of the self-energy, while the reduction of g in the $t \gg J$ limit incorporates the downward renormalization of the coherent part of the propagator by incoherent processes. The scaling $g \sim J$ in the $t \gg J$ limit corresponds to the saturation of the single-vacancy dipole moment at O(1). Note that for a low density of vacancies only states with $k = k_v$ are occupied; at small momentum transfers the dipolar coupling is dominant. The coupling to **m** is suppressed by an extra power of n.

Take the classical limit of Eq. (5.68):

$$\mathscr{H}_{cl} = -q\hat{p}_a(\hat{\Omega} \times \partial_a\hat{\Omega}) + \tfrac{1}{2}\rho(\partial_a\hat{\Omega})^2. \tag{5.73}$$

The dipoles clearly order, $\langle p_a \rangle \neq 0$, leading to a spiral AF phase with $\hat{\Omega} \times \delta\hat{\Omega} = gp_a\rho^{-1}\langle p_a \rangle$, where $\hat{\Omega}$ rotates in the plane perpendicular to the spin direction of $\langle p_a \rangle$ with a pitch along the spatial direction of $\langle p_a \rangle$. The magnitude of the inverse pitch (equivalently the incommensurability wavenumber **Q**) is proportional to the total polarization and therefore to the density of holes $\mathbf{Q} = |\nabla\Omega| = g\rho^{-1}|p_a| \sim n$.

The quantum problem is more subtle since—due to the Pauli principle—$p_a(k)$ cannot be identical for all vacancies. First consider the renormalization of the spin-wave propagator by particle–hole fluctuations as described by bubble diagrams. Only the stiffness constant is modified to lowest order in n since the coupling to **m** vanishes at the zone center. One obtains a renormalized static stiffness $\tilde{\rho} = \rho - g^2\chi_d$ with

$$\chi_d = \frac{1}{6} \sum_{k,a} \sin^2 k_a \int dt \langle \bar{\Psi}_k\hat{\tau}\Psi_k(0)\bar{\Psi}_k\hat{\tau}\Psi_k(t) \rangle, \tag{5.74}$$

the static dipole susceptibility. The instability is signalled by $\tilde{\rho} < 0$:

$$\rho^{-1}g^2\chi_d > 1.$$

For noninteracting particles at zero temperature $\chi_d = 4N_F\langle \sin^2 k_a \rangle$, where N_F is the density of states at E_F and the angular brackets are an average over the Fermi surface. In 2D, $N_F = (\mu_\parallel/\mu_\perp)^{1/2}/2\pi$ and for t/J either large or small $\rho^{-1}g^2\chi_d > 1$ is of order $1 \times (\mu_\parallel/\mu_\perp)^{1/2}$. Thus if the anisotropy $\mu_\parallel/\mu_\perp$ is as large as we expect, the instability occurs at arbitrarily low hole density n. The $n \to 0$ limit is, of course, singular, since the calculated stiffness renormalization only applies at wavenumbers $q^2 \ll k_F^2 \sim n$. Even in this limit the absence of a threshold density is an artifact of 2D and $T = 0$. In the classical limit $\chi_d \sim n/T$, while in 3D at $T = 0$, $\chi_d \sim n^{1/3}$, so that in either case the instability occurs for $n > n_c$.

A state with negative stiffness constant ρ evidently prefers to twist, and one may construct a mean-field theory for such a phase. Assume $\hat{\Omega} \times \nabla\hat{\Omega} = zQ_n \neq 0$ (in the CP′ parameterization one has $\langle \bar{z}_{z+a}z_r \rangle = iQ_a/2$ as the spiral order parameter). The mean field version of Eq. (5.68) reads

$$\mathscr{H}_{MFT} = \sum_k \varepsilon_k n_k - gQ_a(\sin k_a(n_k^+ - n_k^-) + \tfrac{1}{2}\rho Q_a^2, \tag{5.75}$$

where the $n_k^\pm$ are the occupation numbers of spin-z states. Minimizing with respect to Q_a leads to the self-consistency condition

$$Q_a = q\rho \sum_k \sin k_a(n_k^+ - n_k^-), \tag{5.76}$$

with $n_k^\pm = (\exp\{\beta(\varepsilon_{k+} \mp gQ_a \sin k_a - \varepsilon_F)\} + 1)^{-1}$ and ε_F is the chemical potential. Equation (5.76) has a nontrivial solution when $g^2/\chi_d \geq \rho$ (note $\chi_d = g^{-1}\langle \nabla P_a/\nabla Q_a\rangle|_{Q=0}$), hence the $\rho < 0$ instability can be identified with the onset of *incommensurate*, spiral order. For $g^2\chi_d/2 < \rho < g^2\chi_d$ and $n \ll 1$, one finds a spatially uniform fully polarized state with $Q_a \sim n$ along the $(1, 0)$ or $(0, 1)$ directions and a positive stiffness constant. For $\rho < g^2\chi_d/2$ this model suggests phase separation, but this is unphysical, since the long-range Coulomb potential has been neglected. There may, however, also exist an intrinsically disordered phase.

It is clear from Eq. (5.76) above that the dipolar polarization $\langle p_a\rangle \neq 0$ is built up by populating opposite valleys of the Fermi sea with vacancies of opposite pseudo-spin. The ordering is reminiscent of Stoner ferromagnetism, except here there is no net spin polarization. In terms of the fields $\Psi_k^{A,B}$, which create spinless holes on sublattice A or B (shifting k by (π, π) sends $\Psi^A \to \Psi^A$, $\Psi^B \to -\Psi^B$ and is equivalent to rotating the pseudo-spin by π about z), any state with $\langle p_a\rangle \perp \Omega$, as is the case for the spiral spin state, has only off-diagonal pseudospin order: i.e., $\langle\Psi^{+A}\Psi^B\rangle \sim Q_n \sin k_a$, but $\langle\Psi^{+A}\Psi^A\rangle = \langle\Psi^{+B}\Psi^B\rangle$.

The hole wave-function has equal weight on the two sublattices and a fixed phase between them. One can also see that for the spiral phase,

$$\langle c_\sigma^+(r')c_{\pm\sigma}\rangle \sim \exp[iQ(r \mp r')/2]. \tag{5.77}$$

To explore the low-energy excitations of the spiral state with given Q_a semiclassically, we examine the long-wavelength distortions of $\hat{\Omega}$, m^{-1}, and $\mathbf{p}_a$ in Eq. (5.68). Define the linearized staggered magnetization operators $\xi_r[\xi^+, \xi] = 0$ in a rotating frame: $O_r^+\hat{\Omega} = (1 - \xi^+\xi/2, (\xi - \xi^+)/2, (\xi + \xi^+)/2)$ where $\tilde{O}_r$ is a uniform $O(3)$ rotation around z corresponding to the chiral state Q_a. Similarly, the long-wavelength transverse fluctuations of the dipole density are parameterized by

$$O_r^+ p_a = |\langle p_a\rangle|(u_x, u_y, 1 - \tfrac{1}{2}(u_x^2 + u_y^2)), \tag{5.78}$$

with $u_x = \tfrac{1}{2}(\pi^+ + \pi - \xi^+ - \xi)$, $u_y = (\pi - \pi^+)/2i$, and $[\pi_r, \pi_{r'}^+] = \delta_{rr'}$. Introducing the magnetization operator η conjugate to ξ, $[\eta_r, \xi_{r'}^+] = \delta_{rr'}$, substituting into Eq. (5.68), and expanding for small k and Q, one obtains

$$\mathcal{H} = \sum_k \rho Q^2 \pi_k^+ \pi_k + \tfrac{1}{2}\rho Q \cdot k(\pi_k - \pi_{-k}^+)(\xi_k^+ + \xi_{-k}) + \chi^{-1}\eta_k^+\eta_k + \rho k^2 \xi_k^+ \xi_k.$$

$$(5.79)$$

Note that the imaginary part $(\xi^+ - \xi^-)$, decouples from π.

The corresponding branch (1) if the spectrum has the usual spin-wave form $\omega_k^2 = (ck)^2 (c^2 = \rho/\chi)$ with the zero mode being a global rotation of $\hat{\Omega}$ in the plane of the spiral (the phase mode of the spiral). The out-of-plane distortion of $\hat{\Omega}$, the $\xi^+ + \xi$ mode, mixes with the polarization fluctuations π leading to more complex dynamics:

$$\omega_{R,T}^2 = \tfrac{1}{2}(c^2 k^2 + \rho^2 Q^4)\left\{1 \pm \left[1 - \frac{4\rho^2 c^2 Q^4 k_\perp^2}{(c^2 k^2 + \rho^2 Q^4)^2}\right]^{1/2}\right\}, \qquad (5.80)$$

with $k_\perp^2 = k^2 - (k \cdot Q)^2/Q^2$. The upper branch (R) is spin-wave-like for $k \gg Q$; however, mixing with π introduces a gap $\omega_R = \rho Q^2$ at $k = 0$. The transverse fluctuations of the dipole polarization dominate the lower branch, the torsion mode ω_T, which lies entirely on the energy scale $\rho Q^2 \sim n^2 J$. Note, that for $k \parallel Q$, $\omega_T = 0$. This is an artifact of the k, $Q \ll 1$ expansion. Higher-order terms in Eq. (5.71), $O((Qk)^2)$, would induce a stiffness (or diffusivity) term for the polarization and the corrected dispersion relation is obtained by replacing $k_\parallel^2$ by $\tilde{k}_\parallel^2 = k_\parallel^2 + \rho^{-1} D(k^2 - Q^2)^2$.

The remaining zeros of ω_T occur at $k_c = \pm Q_a$ and are associated with uniform rotation of the plane of the spiral in spin space.

The transverse fluctuations of the dipole polarization arise as a collective mode, involving slowly varying perturbations of the Fermi distributions of the vacancies and of the mode structure. The dipole (or torsion) mode is limited to small momentum transfers $q < k_F \sim Q^{1/2}$.

The behavior of the static spin-correlation function with temperature and doping should furnish a useful experimental signature of the spiral state. We expect the incommensurability Q to appear for $T < \varepsilon_F \sim n$ (for $n > n_c$) and remain constant. The spatial direction of Q is difficult to predict without a more realistic Hamiltonian; however, we note that in the $Q = (1, 1)$ state there would be a net interlayer exchange of $O(Q^2)$ (for the LaCuO based material which may make it more favorable). The spin correlations are anisotropic because of the softness of the torsion mode in the $k \parallel Q$ direction. In the classical limit the fluctuations of Ω, parametrized by the polar angles Θ, Φ (with the uniform rotation about z taken out) have energy

$$E \sim [\rho k_\perp^2 + D(k_\parallel^2 - Q^2)^2]\Theta_k^2 + \rho k^2 |\Phi_k|^2, \qquad (5.81)$$

where $q \cdot k_\parallel = q \times k_\parallel = 0$.

The spin-order along Q can disappear for $T \sim JD^{1/2}Q \ll \varepsilon_F$ as the torsion mode "melts," leaving the correlation length $\chi \sim O(n^{-1})$. The presence of domains of different Q and topological defects in the spiral structure, which may be quenched in, will make the correlations more isotropic. However disordered the torsion mode becomes, it cannot reduce the correlation length beyond $Q^{-1} \sim n^{-1}$, since on shorter length scales the spins obey the Heisenberg model, which is ordered. These arguments favoring a spiral state do *not* require long-range order, but only $k_F \ll \xi$, which is satisfied for $n \ll 1$.

Holes also make a potentially important nonhydrodynamic contribution to spin-wave damping via the imaginary part of the dipole susceptibility χ_d. For $k \ll k_F$ one obtains $\Gamma = \text{Im}(\omega/ck) \sim g^2\mu c/\rho v_F$, provided $c < v_F$ or $\Gamma = g^2\mu v_F/\rho c$ for a window around $k \sim \mu c$. For weakly localized holes when the momentum conservation constraint is absent, one finds approximately $\Gamma \sim g^2\mu n/c$. The latter may be important at $k \ll k_F$ for low densities where $c > v_F$ and Γ is otherwise zero.

The torsion mode lying entirely at frequencies $O(n^2 J)$ is easily saturated thermally and is a possible source of the linear T resistivity of the normal state. The resistivity would arise from direct (spin flip) coupling of the holes with the collective torsion mode analogously to itinerant ferromagnets. (Scattering by a spin-diffusion m mode would give $\rho \sim T^2$ or T^l, $l > 2$.)

It has been shown that mobile vacancies in a background with at least short-range AF order have dipolar interactions, which induce their collective polarization, leading to a spiral AF phase (even if the vacancies are not strongly localized). The spiral order implies, at the one-particle level, correlations between the wavenumber and pseudospin which extend throughout the Fermi sea.

We expect that along with the ordered spiral phase there might exist (for larger values of effective coupling constant or higher vacancy density) a disordered state with local spiral twist. Dipolar interactions may also introduce pair correlations and superconductivity in singlet and triplet channels. The competition and coexistence of superconductivity and local spiral order can be explored on the mean-field level using the four-fermion Hamiltonian with dipolar interactions obtained by integrating out the spin waves.

Finally, it appears that the hopping-induced dipole moment of the vacancy persists even in the more realistic two-band model of CuO planes.

5.2.5. Slave bosons

One technique for including aspects of the charge degree of freedom in models mainly concerned with magnetic (spin) degrees of freedom is the slave boson technique [50–54].

The use of slave bosons was first proposed to deal with the infinite-U Anderson model in mixed-valence systems [51, 52, 54]. It introduces a local constraint by extending the Fock space to include auxiliary (slave) boson fields, which track the occupation of the local impurity level. The constraint is implemented by a Lagrange multiplier field λ_i. At the mean-field level, the boson fields b_i and λ_i are constant numbers b and λ; b multiplicatively renormalizes the p–d hopping yerm, while λ additively shifts the atomic levels of the d orbitals (see below).

The principal advantage of the additional boson fields, whose number operator is constrained together with the local fermion variables, is that standard field-theoretic methods can be used. Qualitatively, the splitting of the original fermion quasi-particle into a fermion–boson description allows the dynamics of the quasi-particle hopping accompanied by local spin-polarization changes to be separated into spin motion and spin and charge excitation changes.

We demonstrate in the following section how the use of the slave boson technique allows the description of the quasi-particles in the cuprates in terms of renormalized parameters derived from band structure or cluster calculations. Our treatments closely follows [11, 53].

Consider a model for $La_{2-x}Sr_xCuO_4$ with orbitals on planar Cu and O sites, with the operator $d_{i\sigma}^+$ creating a hole in the $3d_{x^2-y^2}$ orbital with energy ε_d^0, and $c_{i\sigma}^+$ creating a hole in the $2p_{x,y}$ orbital with energy ε_p. A hopping matrix element t_{pd} is also assumed between nearest-neighbor Cu and O sites. We also include Hubbard parameters U_d, U_c for each site, and take $U_d = \infty$, $U_c = 0$. For undoped La_2CuO_4, we assume $\varepsilon_p - \varepsilon_d^0 > 0$, leading to configurations Cu^{2+} and O^{2-}. The holes introduced by doping with strontium occupy oxygen orbitals. The energy difference $D = \varepsilon_p - \varepsilon_d^0$ plays a role analogous to the Coulomb repulsion U in the Hubbard model [55].

Introduce a slave boson at each copper site. As only a single linear combination of the two oxygen orbitals couples to the Cu d orbital (of b_{1g} symmetry [10]), we have a two-band model with energies

$$E_{1,2}(\mathbf{k}) = (\varepsilon_\rho + \varepsilon_d^0 \pm R_k^0)/2,$$
$$R_k^0 = ((\varepsilon_\rho - \varepsilon_d^0)^2 + 16t_{pd}^2\gamma_k^2)^{1/2}, \tag{5.82}$$
$$\gamma_k^2 = \sin^2(k_x/2) + \sin^2(k_y/2),$$

and for one hole per unit cell, the lower band E_1 is filled.

This model is formally equivalent to the Anderson lattice, and we use results from earlier work on large orbital degeneracy (N) expansion [53, 54]. Redefine the kinetic energy (hopping) term to be $t/N^{1/2}$ (e.g., for $N = 2$, $t = \sqrt{2}t_{pd}$). Next assume that the N-fold degenerate d states can accommodate $Q = Nq_0$ holes and set $q_0 = \frac{1}{2}$. The deviation from half-filling, δ, is defined by $H = Q(1 + \delta)$ for the total number of holes per unit cell. The slave bosons are introduced to enforce the constraint

$$b_i^+ b_i + d_i^+ d_i = Q. \tag{5.83}$$

$N = 2$ represents the spin degeneracy. Only if $N = 2$ and $Q = 1$ is the constraint equivalent to $U_d = \infty$.

From the $1/N$ expansion, we know that the lowest order gives mean-field theory (Hartree–Fock approximation) with $\langle b \rangle = b_0 = \sqrt{N}r_0$ and the effective hopping matrix element is $\sigma_0 = (t/\sqrt{N})b_0 = tr_0$. In addition, the position of the d-level is renormalized, $\varepsilon_d^0 \to \varepsilon_d$, and the renormalized band structure obeys Eq. (5.1) with the replacements $t_{pd} \to \sigma_0$ and $\varepsilon_d^0 \to \varepsilon_d$. These bands are the actual quasi-particle bands and one finds the chemical potential by filling the lower band with H/N holes. The Fermi surface contains H holes and the Luttinger theorem is obeyed.

We have to solve for the mean-field parameters r_0, ε_d:

$$r_0^2 + \sum_k u_k^2 f(E_1(\mathbf{k})) = q,$$

$$\varepsilon_d - \varepsilon_d^0 = \frac{t}{r_0} \sum_k u_k v_k \gamma_k f(E_1(\mathbf{k})), \tag{5.84}$$

where

$$\left.\begin{matrix} u_k^2 \\ v_k^2 \end{matrix}\right\} = (1 \pm (\varepsilon_p - \varepsilon_d)/\mathbf{R}_k) \tag{5.85}$$

are the weights of the Cu and O orbital in the lower band.

The limit $4t \ll D$ can be studied analytically in terms of the two dimensionless parameters D/t and δ. For $\delta < 0$ the shift in ε_d can be found perturbatively from

$$r_0^2 = -\frac{\delta}{2} + \sum_k (1 - u_k^2) f((E_1(\mathbf{k})) \tag{5.86}$$

for $t \ll D$, $u_k^2 \approx 1$ and $r_0^2 = -\delta/2$. The bandwidths of the two bands are given by $8r_0^2 t^2/D$, i.e., they are narrowed by a factor $|\delta|$ compared to the bare bandwidth $8t_{pd}^2/D$. This is quite analogous to the band narrowing in the Hubbard model for $U/t \gg 1$.

The physical situation described is a Fermi liquid obeying Luttinger's theorem, in which δ holes in the lower "Hubbard" band correspond to $H = N/2(1 + \delta)$ holes, which have a mass-enhancement $m^*/m \approx 1/\delta$. We also see by inspection of Eq. (5.86) that $r_0 = 0$ at $\delta = 0$, so that the perturbative solution discussed so far disappears for $\delta > 0$. There exists another solution, in which ε_d is renormalized almost up to ε_p and therefore, $\varepsilon_p - \varepsilon_d = 4t^2/D \sum_k \gamma_k^2 f(E_1)$. If $\sigma_0 \ll \varepsilon_d - \varepsilon_p$ (which can be seen to hold), we can expand $\sum_k (1 - u_k^2) f(E_1)$ to be $\gamma_0 4t^2/D^2 \sum_k \gamma_z^2 f(E_1) \gg \gamma_0^2$. For that case we have the solution

$$r_0^2 \approx \frac{\delta}{2} \left(\frac{2t}{D}\right)^2 \sum_k \gamma_k^2 f(E_1(\mathbf{k})) \tag{5.87}$$

and we find the bandwidths are still given by $8\delta t_{pd}^2/D$ as was the case for $\delta < 0$ (see Fig. 5.7).

The discontinuous jump in the position of the bands relative to the $\varepsilon_p, \varepsilon_d^0$ for δ changing sign is due to the fact that the chemical potential must be near ε_p for $\delta > 0$, because the additional holes occupy oxygen sites. The

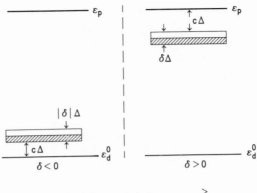

Renormalized Hole Bands for $\delta \lessgtr 0$

$$c = \sum_k \gamma_k^2 f\left(E(k)\right) \approx 1$$

FIG. 5.7.

holes on oxygen can delocalize by virtual hopping into Cu sites with an energy gain of $4t_{pd}^2/D$ from perturbation theory, explaining the narrow band below ε_p (see Fig. 5.7).

Physically, we do not expect band narrowing to continue for arbitrarily small $|\delta|$, as we anticipate an antiferromagnetic (Néel) state at and near half-filling. We should compare J with the delocalization energy $4\delta t_{pd}^2/D$ and expect the Fermi liquid picture to hold for $\delta > t_{pd}^2/D(D + U_c)$.

We note that it is possible to go beyond the present mean-field approximation and include fluctuations in the slave bosons [11].

Extensions to multiband models with several slave bosons can be found in [56–59], which show encouraging agreement with experimental data.

5.3. Non-Fermi-liquid models

5.3.1. The resonant valence bond (RVB) model and its evolution

It was proposed by Anderson [16] at the very beginning of the discovery of superconductivity in the cuprates [7] that magnetic fluctuations were responsible for the pairing. In addition, Anderson suggested [16] that a novel type of ground state consisting of singlet pairs may form a quantum spin liquid which could move on the underlying square lattice by resonant tunneling once the filling moved away from half, in which all the spins are paired. Such a RVB state was proposed by Anderson for a triangular lattice [60] a long time ago. This fascinating proposal was rapidly taken up and expanded upon by many others, in particular, Kivelson, Rokhsar, and Sethna [17].

Because of the very large degeneracy of spin configurations near the half-filled state and also because of the possibility of quantum fluctuations introducing additional novel possibilities for the $s = \frac{1}{2}$ state of the carriers, it took some time before it was demonstrated explicitly by calculation [49] that, in fact, the spin-$\frac{1}{2}$ Heisenberg model in two dimensions has an antiferromagnetic ground state with an effective moment of 0.4 μ_B reduced by transverse quantum fluctuations from the classical value of $\frac{1}{2}\mu_B$.

One of the central points stressed by Anderson [61–63] is the nature of the normal metallic state and, in particular, the temperature dependence of the resistivity, proportional to the temperature T. This empirical law seems to be obeyed by almost all the newly discovered cuprate superconductors with high superconducting transition temperatures.

Furthermore, it was emphasized in the evolution of these ideas [61–63] that the single-band Hubbard model contains the essential physics for the understanding of the low-energy (< 1 eV) behavior of the quasi-particles. The $U \gg 4t$ limit of the Hubbard model is assumed to be the relevant parameter regime and some of the exact results found for the 1D Hubbard model such as charge and spin separation [64] are postulated to also hold in higher dimensions.

A particularly successful application of these ideas was made to experimental determination [65] of the temperature dependence of the Hall effect in [66].

We refer the interested reader to the literature on this topic [61–63] and have also listed several of the proceedings of recent conferences and workshops on the superconducting cuprates [79–83].

5.3.2. *Anyon models and fractional statistics*

The second proposal for a non-Fermi-liquid origin of superconductivity in the cuprates [7] arose from the striking success of the ideas of fractional statistics [67–70] in explaining the fractional quantum Hall effect [71].

In its application to the cuprates, it is argued that spin–charge separation may occur as it does for the one-dimensional Hubbard chain [64] which has been named the "Luttinger liquid." The separate spin- and charge-carrying entities are shown to obey Fermi and Bose statistics, respectively. This break-up of the hole carrier created by doping into separate excitations, "spinons" and "holons," was also an early part of the RVB description of the cuprates.

It was proposed by Kalmeyer and Laughlin [18] that both the chargeless, spin-$\frac{1}{2}$ excitations ("spinons") and the charged, spinless excitations ("holons") obey $\frac{1}{2}$-fractional statistics. The basic analogy is that as charge is fractionalized in the fractional quantum Hall effect, here the spin 1/2 is also fractional.

Although the arguments for such a model to apply for the cuprates are very appealing on aesthetic grounds and found strong resonances with other theorists [72–74], it was soon realized that it made some striking predictions for experimental observation, such as its breaking of parity P (reflection in the plane) and time-reversal symmetry T.

Experiments on muon-spin precession [76] put stringent upper limits on local magnetic fields estimated by Halperin [77] and experimental tests of optical activity and dichroism showed conflicting results. In fact, the most sensitive experiment [78] did not show any effect.

We refer the interested reader to the literature on this topic and have listed several of the proceedings of recent conferences and workshops on the superconducting cuprates [79–83].

5.4. Conclusions

We have considered in this chapter a very tiny subset of the huge number of papers on spin-fluctuation effects in the cuprates. As the major emphasis of the rest of the book is on other mechanisms and our own view of these materials, it seemed necessary for balance to mention this very active field.

The relevant regime for superconductivity starts at small values of the carrier concentration n_c, near the half-filled band, which characterizes the anti-ferromagnetic starting materials La_2CuO_4 and $YBa_2Cu_3O_6$. The central idea of the models we have described is that the motion of the carriers introduced by doping is dominated by the proximity to anti-ferromagnetic order in the CuO_2 planes. The intra-atomic Coulomb repulsion U between two holes on the same site is assumed to remain the dominant energy scale ($U > 4t$). Superconductivity within the Fermi-liquid based models such as the spin-bag and the two-dimensional Hubbard model occurs via the exchange of paramagnons and leads to a d-wave order parameter for the superconducting gap function.

In the non-Fermi-liquid models the superconducting state is not yet expressed in a generally accepted form: in the view of P. W. Anderson inter-layer hopping of the Luttinger liquid is invoked and a gage force arising from the fractional 1/2 statistics of the carriers leads to pairing in R. B. Laughlin's anyon model.

Applicability of the Hubbard model to pairing in cuprates requires an experimental manifestation of the d-wave pairing.

6

INDUCED SUPERCONDUCTIVITY

One can distinguish two types of superconducting states: intrinsic and induced. In the former case the system is characterized by its own intrinsic pairing interaction, and if temperature drops below T_c the material becomes superconducting.

The case of induced superconductivity is entirely different. The system of interest is intrinsically normal, but when it is in contact with another superconductor it becomes superconducting. Charge transfer, which is a consequence of this contact, leads to the appearance of the phenomenon of induced superconductivity.

There are two major channels of such charge transfer. One involves spatial separation of the two subsystems (proximity effect). The other case (two-gap model) corresponds to separation in momentum space, rather than in real space. In the latter case superconductivity can be induced in the normal subsystem by means of interaction with an intermediate field, such as phonons.

Analysis of the properties of the high T_c oxides represents one of the important applications of the phenomenon of induced superconductivity.

6.1. Two-band model

6.1.1. *General description*

Consider a superconductor containing several different groups of electrons occupying distinct quantum states. The most typical example is a material with several overlapping energy bands. One can expect that each band will possess its own energy gap. This means that the density of states of the superconducting pairs contains several peaks. Of course if the energy gap were defined as the smallest quantum of energy that can be absorbed by the material, then only the smallest gap of the system would satisfy this definition. Thus to avoid misunderstanding, when talking about multigap structure of a spectrum we will mean explicitly the aforementioned multipeak property of the density of states.

The magnitudes of these gaps will differ according to the variations in the densities of states and coupling constants across the bands.

The two-band model was first considered in [1, 2]. A general treatment based on quantum field-theoretic techniques, together with an analysis of various thermodynamic, magnetic, and transport properties, was given in [3–5].

Each band contains its own set of Cooper pairs. Since, generally speaking, $p_{Fi} \neq p_{Fk}$ (here p_{Fi} and p_{Fk} are the Fermi momenta for different bands), there is no significant pairing of electrons belonging to different energy bands. This does not mean, however, that the pairing within each band is completely insensitive to the presence of the other. On the contrary, a peculiar interband interaction and the appearance of nonlocal coupling constants are fundamental properties of the multiband model.

Consider two electrons belonging to band i. They exchange phonons and form a pair as a result. There exists two pairing scenarios. In one of them, the first electron emits a virtual phonon and makes a transition into a state within the same energy band. The second electron absorbs the phonon and also remains in the same energy band, forming a bound pair with the first one. This is the usual pairing picture of the BCS theory, described by a coupling constant λ_i. However, the presence of the other energy band gives rise to an additional channel. Namely, the first electron, originally located in the i band, can emit a virtual phonon and make a transition into the k band. The phonon is absorbed by the second electron, which also is scattered into the k band, where it pairs up with the first electron.

As we know, there is no energy conservation requirement for single virtual transitions; such conservation, however, must hold for the initial and final states. In our case this criterion is met. Indeed, the initial and final states correspond to particles on the Fermi surface. Note that, in addition, the initial and final total momenta are equal to zero.

Thus the initial state had two electrons in the i band, while the final state finds a pair in the k band. Interband charge transfer processes are described by nondiagonal coupling constants λ_{ik}, and because of them the system is characterized by a single critical temperature (see below). Otherwise, each set of electrons would have its own T_c.

6.1.2. Critical temperature

We introduce self-energy parts $\Delta_i(\mathbf{p}, \omega_n)$ which describe electron pairing in the ith band. They satisfy the system or equations shown in Fig. 6.1, or, in analytical form

$$\Delta_i(\mathbf{p}, \omega_n) = \sum_{l=1}^{n} \sum_{\omega_{n'}} \int \lambda_{il} D(\mathbf{p} - \mathbf{p}', \omega_n - \omega_{n'}) F_{ll}^+(\mathbf{p}', \omega_{n'}), \qquad (6.1)$$

where

$$F_{ll}^+(\mathbf{p}, \omega_n) = \frac{\Delta_l}{\omega_n^2 + \xi^2 + \Delta_l^2}.$$

D is the phonon Green's function; λ_{il} describes the transition of an electron

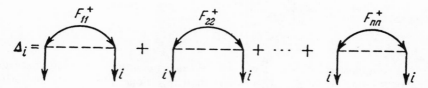

FIG. 6.1. Pairing in a multiband case.

pair from the ith to the lth band accompanied by phonon absorption (or emission).

Let us look at the case of weak coupling. Introducing the effective coupling constants in the usual way and summing over $\omega_{n'}$, we write (6.1) in the form

$$\Delta_i = \sum_{l=1}^{n} g_{il} v_l \Delta_l \int d\xi_l \frac{\tanh(\varepsilon_l/2T)}{\varepsilon_l},$$

$$\varepsilon_l = (\xi_l^2 + |\Delta_l|^2)^{1/2}. \tag{6.2}$$

One can see from Eq. (6.2) that the gaps in different bands are linearly related to each other. Therefore all of them appear at the same temperature T_c which can be calculated from Eq. (6.2). In the two-gap model the critical temperature turns out to be

$$T_c = 1.14\tilde{\Omega} e^{-1/\tilde{\lambda}}, \qquad \tilde{\lambda} = \tfrac{1}{2}[\lambda_a + \lambda_b + \sqrt{(\lambda_b - \lambda_a)^2 + 4\lambda_{ab}\lambda_{ba}}]. \tag{6.3}$$

This result can be easily obtained from Eq. (6.2) $(i, l = 1, 2)$ taken at $T = T_c$.

In the usual theory of superconductivity, in the weak-coupling approximation the energy gap is related to T_c by the universal relation $2\varepsilon(0) = 3.52 T_c$ (see Section 2.5). In a multiband model with a large number of coupling constants there is no such universal relation between $\varepsilon(0)$ and T_c.

The solution of Eq. (6.1) at $T = 0$ K can be obtained by analogy with the method used in [6]; it can be sought in the form $\Delta_i = 2\tilde{\Omega}\phi_i \exp(-\alpha)$, $\phi_i = QX_i$ where X_i is the normalized solution of the following system of linear equations:

$$\phi_i = \alpha \sum_l \lambda_{il} \phi_l \tag{6.4}$$

corresponding to the smallest eigenvalue α. The coefficient Q is equal to:

$$Q = \exp\left[-\sum_i v_i |X_i|^2 \ln|X_i| \Big/ \sum_i v_i |X_i|^2 \right],$$

$$X_1 = \lambda_{12} S^{-1}, \qquad X_2 = (\tilde{\lambda} - \lambda_{11}) S^{-1}, \tag{6.5}$$

where

$$S = [\lambda_{12}^2 + (\tilde{\lambda} - \lambda_{11})^2]^{1/2}.$$

From these results one can show [5] that one of the gaps always exceeds the BCS value, while the other is less than this value. Indeed, we can write

$$\frac{2\varepsilon_1(0)}{T_c} = 3.52\, e^{A_1}, \qquad \frac{2\varepsilon_2(0)}{T_c} = 3.52\, e^{A_2}, \qquad (6.6)$$

where

$$A_1 = v_2\tau^2 R \ln(\tau^{-1}), \qquad A_2 = -v_1 R \ln(\tau^{-1}),$$
$$R = (v_1 + v_2\tau^2)^{-1}, \qquad \tau = (\tilde{\lambda} - \lambda_{11})/\lambda_{12},$$

or

$$\tau = \tfrac{1}{2}(\lambda_{22} - \lambda_{11} + [(\lambda_{22} - \lambda_{11})^2 + 4\lambda_{12}\lambda_{21}]^{1/2})\lambda_{12}^{-1}.$$

One can see that regardless of the value of τ we always have $A_1 A_2 < 0$. Therefore one can state the following theorem: One of the gaps in a two-gap system is always smaller than the BCS value, whereas the other gap is greater than this value. Note that deviation from the BCS behavior here is caused not by strong coupling (see Chapter 2) but by the presence of a multigap structure.

6.1.3. Two-gap spectrum and properties of superconductors

The temperature dependences of the thermodynamic, electromagnetic, and transport parameters also differ from those obtained in the usual one-gap BCS theory.

For example, for $T \to 0$ the heat capacity is the sum of two exponents; the band with the smaller superconducting gap makes the dominant contribution, and as a result $C_{el; T \to 0}$ decreases slower than in the one-band case.

The same is true for the surface resistance R_s. R_s is a sum of two terms, and in the low-temperature region it is a sum of two exponents:

$$R_s = a\, e^{-\varepsilon_1/T} + b\, e^{-\varepsilon_2/T}. \qquad (6.7)$$

If $\varepsilon_1 \gg \varepsilon_2$, then at $T \to 0$ the surface resistance is dominated by the smaller gap ε_2.

The electrodynamics of two-band superconductors turns out to be very peculiar. For example, one can have situations in which $\xi_2 \ll \delta \ll \xi_1$ (δ is the penetration depth, ξ_1 and ξ_2 are the coherence lengths in the two bands). In this case a single sample will contain two groups of electrons, one of which is Pippard-like, while the other is London-like.

The presence of overlapping bands is typical for many superconductors. Nevertheless, it was not possible to observe multigap structure in conventional superconductors. This is due to their large coherence lengths: the inequality $l \ll \xi_0$ (l is the mean free path), which holds for most conventional superconductors, leads to averaging [7] because of interband scattering. As a result, the multigap structure is washed out and the one-gap model is applicable. A two-gap spectrum was first observed in the exotic system $SrTiO_3$ [8].

The situation in the high T_c oxides is entirely different. The coherence length is small, and there exists a real opportunity to observe a multigap structure. This subject will be discussed in detail in Chapter 7.

6.1.4. *Induced two-band superconductivity*

Let us consider the two-band model and the special case of $\lambda_{22} = 0$, i.e., let us suppose that one of the bands is intrinsically normal. Below the critical temperature, both bands will be in the superconducting state. In other words, interband transitions will induce superconductivity in the second band. The critical temperature is given by (6.3) with

$$\tilde{\lambda} = \tfrac{1}{2}[\lambda_{11} + (\lambda_{11}^2 + 4\lambda_{12}\lambda_{21})^{1/2}]. \tag{6.8}$$

In the present case the bands are not spatially separated. Instead, near the Fermi surface the states belonging to different bands are separated in momentum space. The induced superconducting state arises thanks to phonon exchange. Note that other excitations such as excitons, plasmons, etc. are also capable of providing interband transitions.

It is important to note that $\tilde{\lambda} > \lambda_{11}$, which is due to an effective increase in the phonon phase space. This means that the presence of the second band is favorable for superconductivity. As a result of charge transfer, the second band acquires an induced energy gap which depends on the constant λ_{21}.

As remarked above, the ratios $2\varepsilon_1(0)/T_c$ and $2\varepsilon_2(0)/T_c$ differ from the value $a = 3.52$ of the BCS theory. Namely, $2\varepsilon_1(0)/T_c > a_{BCS} = 3.52$ and $2\varepsilon_2(0)/T_c < a_{BCS}$. In the case of induced two-gap superconductivity, $\lambda_{11} \gg \lambda_{21}$ and $\lambda_{22} = 0$. Then we find that the ratio $\tau = \lambda_{21}/\lambda_{11} \ll 1$, and the values of the energy gaps may differ noticeably from each other.

Now consider the case of strong coupling. If we assume that $\lambda_{11} \gg \lambda_{21}$, we can write

$$\Delta_1(\omega_n)Z_1 \cong \lambda_{11} \sum_{\omega_{n'}} \kappa_{\omega_n, \omega_{n'}} \Delta_1(\omega_{n'}), \tag{6.9}$$

$$\Delta_2(\omega_n) \cong \lambda_{21} \sum_{\omega_{n'}} \kappa_{\omega_n, \omega_{n'}} \Delta_1(\omega_{n'}), \tag{6.9a}$$

where

$$\kappa_{\omega_n, \omega_{n'}} = \tilde{\Omega}^2 [\tilde{\Omega}^2 + (\omega_n - \omega_{n'})^2]^{-1} (\omega_{n'}^2 + \Delta_1^2)^{-1/2}. \tag{6.9b}$$

The energy gap is the solution of the equation $\omega = \Delta(-i\omega)$ where $\Delta(z)$ is the analytical continuation of the function $\Delta(\omega_n)$. We obtain for the ratio of the gaps [5]:

$$\tau = \frac{\varepsilon_2(0)}{\varepsilon_1(0)} \cong \frac{\lambda_{21}(1 + \lambda_{11})}{\lambda_{11}}. \tag{6.10}$$

The presence of strong coupling leads to the appearance of the factor $(1 + \lambda_{11})$. We see that strong coupling terms to decrease the relative difference in the values of the gaps.

Two-band induced superconductivity is caused by phonon-mediated exchange, but does not require a spatial separation of the intrinsically superconducting and normal subsystems. Another type of induced superconductivity, the proximity effect, is based on such a separation and will be considered in the next section.

6.2. Proximity effect

6.2.1. Proximity "sandwich"

The second type of induced superconductivity, so-called proximity effect, does occur in conventional superconductors; it was discovered in 1961 [9] and involves spatially separated normal and superconducting subsystems. The simplest type of proximity system is shown schematically in Fig. 6.2. A superconducting state is induced in the normal film N under the influence of the neighboring superconducting film S. Experimentally one finds that film N begins to exhibit the Meissner effect [10].

An important parameter characterizing the proximity system is the coherence length ξ_N in the normal film. This quantity is temperature-dependent and equals $\xi_N = \hbar v_F / kT$ [11]. Thus lowering the temperature

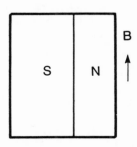

FIG. 6.2. Proximity system.

benefits the proximity effect. It is for this reason that the largest-scale observation of induced superconductivity to date (up to 2×10^5 A) was accomplished at ultralow temperatures in NbTe–Cu(Ag) systems [12].

There exist different approaches to describing the proximity effect; among them, two are particularly prominent. One is based on the Ginzburg–Landau theory and makes use of the boundary conditions at the S–N interface (the "Orsay school," see [13]). This approach directly takes into account the spatial variation of the order parameter $\Delta(\mathbf{r})$.

The other approach is the tunneling model developed by McMillan [14]. The superconducting state in the N film arises thanks to Cooper pair tunneling from the superconducting film into the normal one. In the tunneling model it is assumed that the order parameters in the S and N films are constant throughout the films. In other words, it is assumed, for example, that $L_N \ll \xi_N$. An attractive feature of this model, in addition to its elegance and simplicity, is that it is applicable not only near T_c but at any temperature.

6.2.2. Critical temperature. Induced energy gap

Let us calculate the critical temperature of an S–N proximity sandwich within the framework of the McMillan model. We focus on this model because it is relevant to the intrinsic proximity effect (see Section 6.3) which is an important feature of the high T_c oxides.

We will employ the method of thermodynamic Green's functions [15]. Let us introduce the self-energy parts Σ_2^α and Σ_2^β describing pairing in the films; they satisfy the following equations:

$$\Sigma_2^\alpha = \Sigma_{2,\mathrm{ph}}^\alpha + \tilde{T}^2 \int d\mathbf{p}' \, F^\beta(\mathbf{p}', \omega_n),$$

$$\Sigma_\alpha^\beta = \tilde{T}^2 \int d\mathbf{p}' \, F^\alpha(\mathbf{p}', \omega_n), \qquad (6.11)$$

$$\Sigma_{2\mathrm{ph}} = T \sum_{\omega_{n'}} \int d\mathbf{p}' \, g_\alpha^2(\mathbf{p}, \mathbf{p}') D(\omega_n - \omega_{n'}, \Omega(q)) F^\alpha(\mathbf{p}', \omega_{n'}).$$

Here D is the phonon Green's function (2.37), and F^α and F^β are the anomalous Green's functions:

$$F^\alpha(\omega_n, \mathbf{p}) = -\frac{\Sigma_2^\alpha(\omega_n, \mathbf{p})}{[\omega_n^2 Z_\alpha^2 + \xi_\alpha^2(\mathbf{p}) + \Sigma_2^{(\alpha)^2}(\omega_n, \mathbf{p})]}. \qquad (6.12)$$

A similar equation holds for F^β. Here ξ_α is the energy of an ordinary electron referred to the Fermi level, Z is the renormalization function, and $\tilde{T}$ is the tunneling matrix element [16, 17].

Equations (6.11) are written for the case when the existence of the pair condensate in the β film is totally due to the proximity effect. In the weak coupling approximation the renormalization functions are equal to

$$Z_\alpha(\omega_n) = 1 + \frac{\Gamma^{\alpha\beta}}{|\omega_n|},$$

$$Z_\beta(\omega_n) = 1 + \frac{\Gamma^{\beta\alpha}}{|\omega_n|}. \tag{6.13}$$

Here

$$\Gamma^{\alpha\beta} = \pi \tilde{T}^2 v_\beta V_\beta,$$

$$\Gamma^{\beta\alpha} = \pi \tilde{T}^2 v_\alpha V_\alpha. \tag{6.14}$$

The parameters $\Gamma^{\alpha\beta}$ and $\Gamma^{\beta\alpha}$ were introduced in [14] and can be written in the form

$$\Gamma^{\beta\alpha} = \frac{V_{F_\perp} \sigma}{2BL_\beta}, \tag{6.14a}$$

where $V_{F_\perp}$ is the Fermi velocity, σ is the barrier penetration probability, and B is a function of the ratio of the mean free path to the film thickness. If $T = T_c$, we should put $\Sigma^{\alpha(\beta)} = 0$ in the denominator of (6.12).

Making use of Eqs. (6.11–6.13), we arrive, after some manipulations, at an equation for T_c:

$$\ln \frac{T_c}{T_c'} = -\frac{\Gamma_{\alpha\beta}}{\Gamma} \frac{1}{\lambda_\alpha} \int d\Omega \, g(\Omega)$$

$$\times \left\{ \left[\psi\left(\frac{1}{2} + \frac{\Gamma}{2\pi T_c}\right) - \psi\left(\frac{1}{2}\right) \right] \frac{\Omega^2}{\Omega^2 + \Gamma^2} + \frac{\Gamma^2}{\Omega^2 + \Gamma^2} \ln \frac{2\Omega\gamma}{\pi T_c} \right\}. \tag{6.15}$$

Here T_c^α is the critical temperature in an isolated α film, and ψ is the digamma function. It should be noted that in general, $g(\Omega) \neq g^\alpha(\Omega)$; in other words, the presence of the interface may alter the function $g(\Omega)$. We will not consider this effect here. The function $g(\Omega)$ is defined on p. 17.

Equation (6.15) allows us to evaluate T_c for any proximity system and is valid for any relation between Γ and T_c^α.

Let us consider some special cases. Assume that $\Gamma \gg T_c$. Making use of the asymptotic form of the digamma function, we find, after simple substitutions, that

$$T_c = T_c^\alpha \left(\frac{\pi T_c^\alpha}{2\langle u \rangle \gamma} \right)^\rho, \tag{6.16}$$

where

$$\rho = \frac{\Gamma_{\alpha\beta}}{\Gamma_{\beta\alpha}}, \qquad \Gamma = \Gamma_{\alpha\beta} + \Gamma_{\beta\alpha} = \Gamma_{\beta\alpha}(1 + \rho)$$

or

$$\rho = \left(\frac{v_\beta}{v_\alpha}\right)\left(\frac{L_\beta}{L_\alpha}\right); \qquad u = \Gamma^{1-\delta}\Omega^\delta; \qquad \delta = \Gamma^2(\Omega^2 + \Gamma^2)^{-1} \qquad (6.16a)$$

The mean value is to be understood in the following sense:

$$\ln\frac{\langle u(\Omega)\rangle}{T_c} = \frac{1}{\lambda_\alpha}\int d\Omega\, g_\alpha(\Omega)\ln\frac{u(\Omega)}{T_c}. \qquad (6.16b)$$

Equation (6.16) is valid for any relation between Γ and $\langle\Omega\rangle$ (keeping in mind that $\Gamma \gg T_c$). If $\Gamma \ll \langle\Omega\rangle$, we arrive at the expression obtained in [14]:

$$T_c = T_c^\alpha\left(\frac{\pi T_c^\alpha}{2\Gamma\gamma}\right)^\rho \qquad (6.17)$$

In the opposite limit, the so-called Cooper case [18], we obtain the result

$$T_c = T_c^\alpha\left(\frac{\pi T_c^\alpha}{2\langle\Omega\rangle\gamma}\right)^\rho \qquad (6.18)$$

Finally, if $\Gamma \le T_c$, $\langle\Omega\rangle \gg T_c$, and $\rho \ll 1$, we find

$$T_c = T_c^\alpha \exp(-F), \qquad (6.19)$$

where

$$F = (\Gamma_{\alpha\beta}/\Gamma)\left[\psi\left(\frac{1}{2} + \frac{\Gamma}{2\pi T_c}\right) - \psi(\tfrac{1}{2})\right]. \qquad (6.19a)$$

Another interesting quantity to determine is the induced energy gap $\varepsilon_\beta(0)$. It can be found from Eq. (6.11) and is given by the simple expression [14]

$$\varepsilon_\beta(0) = \Gamma^{\alpha\beta}. \qquad (6.20)$$

Figure 6.3 displays the density of states of the superconducting electrons evaluated in [14]. Indeed, the density of states is characterized by two distinct peaks.

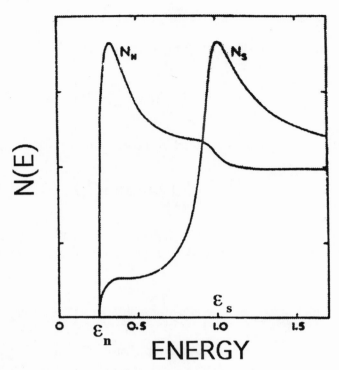

FIG. 6.3. Densities of states for S–N system ("two-gap structure").

6.3. Proximity effect vs. the two-gap model

At this point we may compare the results found for the proximity system with those of the two-gap model. There is a profound analogy between these two situations. In both cases we are dealing with induced superconductivity (we are assuming that in the second band, as in the N film, there is no intrinsic pairing). In both cases an induced energy gap appears. However, the mechanisms giving rise to induced superconductivity are very different. In the two-band model the systems are "separated" in momentum space, and the second band acquires an order parameter thanks to phonon exchange. The phase space for phonons is effectively increased. In the proximity effect, on the other hand, the systems are spatially separated, and superconductivity is induced by the tunneling of Cooper pairs.

In both cases we end up with an induced energy gap, but the physical differences manifest themselves in the behavior of the critical temperature. The two-band picture turns out to be favorable for T_c which becomes higher than in the single-band case (see Section 6.1.4). On the contrary, the proximity effect depresses T_c. The critical temperature of a proximity sandwich is lower than that of an isolated S film. This can be seen directly from Eqs. (6.16) and (6.18).

The physics of proximity systems is an interesting and broad area. For example, the study of Josephson junctions S–N–I–S and S–N–S is concerned with Josephson tunneling in proximity systems. Other active areas include the electrodynamics of proximity systems, as well as the properties of S_α–S_β ($T_c^\alpha \neq T_c^\beta$) contacts. These questions, however, are beyond the scope of the present book.

6.4. Layered systems. General case

In this section we are concerned with a general picture of induced super-conductivity [19, 20]. We consider the case of two conducting subsystems, one of them is intrinsically superconducting and the other is intrinsically normal. Charge transfer leads to the situation where both subsystems become superconducting below some critical temperature. This means that the intrinsically normal subsystem is characterized by an induced superconducting state. One can introduce the general Hamiltonian which describes this phenomenon.

Some related phenomena were described above, e.g., the proximity effect between a superconducting and a normal (S–N) system. The proximity effect assumes a spatial, macroscopic separation between the S and N subsystem, for example, a sandwich consisting of two films separated by an interface. Another example is the two-band case, when one of the bands is intrinsically superconducting and the quantum states of the other band form an intrinsically normal system. In this case we do not have spatial separation between the bands, but the states are separated in momentum space. Contrary to the proximity effect, induced superconductivity in this case requires the exchange of some excitations.

If we study a layered system, like the high T_c cuprates or conventional superlattices with very small layer thicknesses, then we are also dealing with induced superconductivity since some of the layers can be intrinsically normal.

6.4.1. Hamiltonian. General equations

Consider a system which consists of two subsystems that contain carriers. One of them (α) is intrinsically superconducting, with T_c equal to T_c^α. The second subsystem (β) is intrinsically normal, i.e., T_c^β is equal to zero. In addition, there is charge transfer between these two subsystems. The α–β system is described by the total Hamiltonian:

$$\hat{H} = H_0^\alpha + H_0^\beta + \sum_{\chi,\kappa,q} g_{\chi\kappa q}^{\alpha\beta} a_\chi^{\alpha+} a_\kappa^\beta b_q + \sum_{\chi,\kappa} T_{\chi\kappa}^{\alpha\beta} a_\chi^{\alpha+} a_\kappa^\beta + \text{c.c.} \qquad (6.21)$$

H_0^α and H_0^β are the terms describing the isolated α and β subsystems; a, a^+, b are the carrier and phonon amplitudes in the second quantization representation; χ, κ, q are the sets of quantum numbers describing the electronic states for the α and β subsystems and the phonons, respectively; $g_{\chi\kappa q}^{\alpha\beta}$ is the electron–phonon matrix element for the $\alpha \to \beta$ transition; and $T_{\chi,\kappa}$ is the tunneling matrix element. The third term describes the phonon-mediated transitions, and the fourth term represents the tunneling Hamiltonian [21]. The third term describes the process which is analogous to inelastic tunneling.

The Hamiltonian H_0^α contains the interaction which lead to the pairing. For example H_0^α contains the term

$$\sum g_{\chi\chi'}^\alpha a_\chi^{\alpha+} a_{\chi'}^\alpha b_q \tag{6.22}$$

if the intrinsic superconductivity for the α subsystem is due to the usual phonon superconductivity. It is important to stress that the model is general and is not restricted only to phonon interactions. The pairing in the α system can be caused by nonphonon mechanisms just as well. The presence of phonons in the last term also does not exclude a more general treatment in which the inelastic charge transfer is caused by other excitations. In addition, even if the pairing interaction in the α system is not phononic, we always have phonons which are able to provide this charge transfer.

Note also that the terms H_0^α and H_0^β also could contain scattering by impurities, magnetic as well as nonmagnetic. It turns out that the presence of magnetic impurities in the intrinsically normal group leads to a particularly interesting situation. We consider this case separately (see below, Section 6.4.4).

The equation for the self-energy parts Δ_α and Δ_β describing pairing interaction can be written in diagrammatic form (Fig. 6.4) or in analytical form (at $T = T_c$):

$$\Delta^\alpha(\omega_n)Z^\alpha(\omega_n) = \pi T \sum_{n'} (\lambda_\alpha D_{nn'} - \tilde{\mu}) \frac{\Delta^\alpha(\omega_{n'})}{|\omega_{n'}|} + \sum_{n'} \lambda_{\alpha\beta} D_{nn'} \frac{\Delta^\beta(\omega_{n'})}{|\omega_{n'}|}$$

$$+ \Gamma^{\alpha\beta} \frac{\Delta^\beta(\omega_n)}{|\omega_n|}, \tag{6.23}$$

$$\Delta^\beta(\omega_n)Z^\beta(\omega_n) = \pi T \sum_{n'} \lambda_{\beta\alpha} D_{nn'} \frac{\Delta^\alpha(\omega_{n'})}{|\omega_{n'}|} + \Gamma^{\beta\alpha} \frac{\Delta^\alpha(\omega_n)}{|\omega_n|}, \tag{6.24}$$

$$Z^\alpha(\omega_n) = 1 + (\lambda_\alpha + \lambda_{\alpha\beta}) \sum_{n'} D_{nn'} \frac{\omega_{n'}}{|\omega_{n'}|} + \frac{\Gamma^{\alpha\beta}}{|\omega_n|}, \tag{6.25}$$

$$Z^\beta(\omega_n) = 1 + \lambda_{\beta\alpha} \sum_{n'} D_{nn'} \frac{\omega_{n'}}{|\omega_{n'}|} + \frac{\Gamma^{\beta\alpha}}{|\omega|}. \tag{6.26}$$

The equations (6.23–6.26) are the basic equations of the theory. Let us discuss the meaning of each of the terms. If we consider only the first terms in the

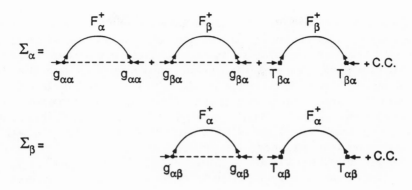

FIG. 6.4.

right sides of the equations for Δ_α and Z^α, then we have the usual Eliashberg equation for the single gap picture. Therefore, the system of two equations represents the generalization of the usual Eliashberg equation (2.37) for the two subsystems with charge transfer between them.

The treatment in a weak-coupling case can easily be modified to consider a nonphonon mechanism. Indeed, in this case (then $\pi T_c \ll \Omega$) one can put $D \approx 1$ and introduce the cut-off at the energy $\approx \Omega$; it corresponds to the usual BCS approximation. Then the nonphonon mechanism can be analyzed by similar formalism with Ω replaced by characteristic excitation energy, and λ_α by the constant describing coupling to these excitations.

It is important to note that Eq. (6.24) for Δ_β does not contain a term similar to the first term in the equation for Δ_α. This reflects the fact that the β subsystem is intrinsically normal ($\lambda_\beta = 0$) and its superconducting state is induced and is described by the terms on the right-hand side of the equation (Eq. (6.24)). The third term on the right-hand side of the equation for Δ_α describes the intrinsic proximity effect. We call this channel an intrinsic proximity effect, stressing that we are talking about a phenomenon which occurs on the scale of a unit cell. This description is analogous to the McMillan tunneling proximity model (see Section 6.2). This term as well as the second term of the following equation for Δ_β describe the tunneling of Cooper pairs between subsystems. This process is possibly only because of the spatial separation of the α and β subsystems.

The second term in the equation for Δ_α, and the first term in the equation for Δ_β, represent the phonon-mediated induced pairing. More specifically, one carrier from the α subsystem radiates a phonon and makes a transition to β. Another carrier absorbs this phonon and as a result they form a Cooper pair in the β subsystem. If the α and β subsystems are spatially separated, then $\lambda_{\alpha\beta}$ contains the same matrix element as inelastic tunneling (see e.g. [22, 23]). However, because the pairing involves the exchange of virtual (not real) phonons, the process described by this term does not represent a real inelastic process (there is no radiation of real phonons). As a result, all initial and final states of the particles are on the Fermi surface so that energy is conserved in the whole process. Similar terms play an important role, even in the case when the α and β subsystems are not spatially separated, for example, in the case of two energy bands. In this case, we have separation only in momentum space, not real space. Then the coupling constant $\lambda_{\alpha\beta}$ contains matrix elements which correspond not to inelastic tunneling but to phonon-mediated transitions between two energy bands. This channel is similar to the two-band model (see Section 6.1).

One can see directly from these equations (6.23–6.26) that each subsystem has its own order parameter, but, nevertheless, the charge transfer leads to linear coupling between them near T_c. Therefore, if Δ_α becomes equal to zero at some T_c, then Δ_β must also be zero at the same T_c. Therefore, our system has two order parameters but only one transition temperature.

The presence of two separate order parameters leads to the appearance of a "two-gap" structure. This means, of course, that the density of states has two peaks which can be determined spectroscopically (see Section 6.1.1). As we know, the energy gap is the minimum energy below which the density of states is equal to zero. According to this definition, our system has a single energy gap which is equal to the smallest (ε_β) one. In the following, we will use the two-gap picture to refer to the two peaks in the density of states.

6.4.2. Critical temperature

6.4.2.1. Weak coupling

Let us consider the case when $\lambda_\alpha \ll 1$. This corresponds to the inequality $T_c \ll \tilde{\Omega}/2\pi$ (see Section 2.3). In this weak-coupling approximation, one can neglect the second term in the renormalization function, Z^α, and a similar term in Z^β; one can neglect also the dependence of the phonon Green's function on ω_n [24]. As a result, we obtain the following equations for the unrenormalized self-energy parts $\Sigma_i(\omega_n) = \Delta_i(\omega_n)Z_i(\omega_n)$ ($i = \alpha, \beta$):

$$\Sigma_\alpha(\omega_n) = \lambda_\alpha \sum_{n'} D_{n'} F_{n'}^\alpha + \lambda_{\alpha\beta} \sum_{n'} D_{n'} F_{n'}^\beta + \gamma_{\alpha\beta} F_n^\beta,$$

$$\Sigma_\beta(\omega_n) = \lambda_{\beta\alpha} \sum_{n'} D_{n'} F_{n'}^\alpha + \gamma_{\beta\alpha} F_n^\alpha,$$

(6.27)

where

$$D_n = \frac{\Omega^2}{\Omega^2 + \omega_n^2} = \frac{v^2}{v^2 + (n + \frac{1}{2})^2},$$

$$F_n^i = \pi T \frac{\Sigma^i(\omega_n)}{|\omega_n| + \Gamma_{ik}} = \frac{\Sigma_n^i}{|2n + 1| + \gamma_{ik}},$$

(6.28)

$$\gamma_{ik} = \frac{\Gamma_{ik}}{\pi T_c}, \qquad i, k \equiv \{\alpha, \beta\}, \qquad i \neq k, \qquad v = \frac{\Omega}{2\pi T_c}.$$

We then substitute Δ_β into the equation for Δ_α and arrive at the following equation:

$$\Sigma_\alpha^n = \lambda_\alpha \sum_{n'} D_{n'} F_{n'} + \lambda_{\alpha\beta} \lambda_{\beta\alpha} \sum_{n',n''} D_{n'} D_{n''} (|2n' + 1| + \gamma_{\beta\alpha})^{-1} F_{n''}^\alpha$$

$$+ \lambda_{\alpha\beta} \gamma_{\beta\alpha} \sum_{n'} D_{n'} (|2n' + 1| + \gamma_{\beta\alpha})^{-1} F_{n'}^\alpha + \gamma_{\beta\alpha} \lambda_{\beta\alpha} (|2n + 1| + \gamma_{\beta\alpha})^{-1} \sum_{n'} D_{n'} F_{n'}^\alpha$$

$$+ \frac{\gamma_{\alpha\beta} \gamma_\alpha}{(|2n + 1| + \gamma_{\beta\alpha})(|2n + 1| + \gamma_{\alpha\beta})} \Sigma_\alpha^n. \tag{6.29}$$

Each of the terms in Eq. (6.29) describes some definite process (see Fig. 6.5). The first term corresponds to an intrinsic pairing for the α group of carriers. All other terms describe different two-step processes. The second term is two consecutive $\alpha \to \beta \to \alpha$ phonon-mediated transitions (two-band channel). The last term represents two tunneling transitions (intrinsic proximity channel). A very important feature of this result is the appearance of a mixed channel described by the third and the fourth terms. For example, the fourth term corresponds to the phonon-mediated transition $\alpha \to \beta$ followed by proximity $\beta \to \alpha$ tunneling. Consider two of the most interesting cases, which correspond to different strengths of the proximity effect. This strength is described by the parameter $\gamma = \Gamma/(\pi T_c)$, $\Gamma = \Gamma_{\alpha\beta} + \Gamma_{\beta\alpha}$. One of these cases occurs when $\Gamma \gg \Omega$ (in addition $\Omega \gg 2\pi T_c$, see above). This case corresponds to a large tunneling probability and is known as the Cooper limit [14, 18]. The opposite limit is characterized by the inequality $\gamma \ll 1$.

Consider first the case of a small tunneling probability ($\gamma \ll 1$). For example, a small tunneling probability between the layers can be due to a large separation between the layers. In this case, the critical temperature is

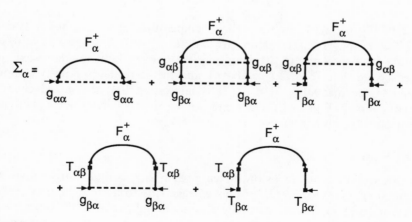

FIG. 6.5.

described by the expression

$$T_c = T_c^0 \exp(f),\tag{6.30}$$

$$f = \frac{\lambda_{\alpha\beta}\lambda_{\beta\alpha}}{\lambda_\alpha^3} + \frac{\pi}{8}\frac{\lambda_{\alpha\beta}}{\lambda_\alpha}\gamma_{\beta\alpha}^0 - \frac{\pi}{16}\gamma_{\alpha\beta}^0.\tag{6.31}$$

Here $\gamma_{ik}^0 = \Gamma_{ik}/\pi T_c$, and T_c^0 is a critical temperature for an isolated α group.

If in these expressions, we put γ_{ik} equal to zero we obtain the expression that corresponds to the usual two-band picture. Indeed, by using the expression for T_c for the two-band model and assuming that $\lambda_\beta = 0$, $\lambda_{\alpha\beta} \ll \lambda_\alpha$, $\lambda_{\beta\alpha} \ll \lambda_\alpha$, one can obtain exactly the first term on the right-hand side of Eq. (6.31). We will discuss some features of this case below (see Section 6.4.2.4).

Let us now consider the opposite limit (Cooper case), when $\Gamma \gg \tilde{\Omega}$. In fact, experimentally one can go from the weak proximity case to the strong proximity case by varying the distance between the layers. Then we obtain

$$T_c = T_c^0\left(\frac{T_c^0}{1.14\tilde{\Omega}}\right)^\beta,\tag{6.32}$$

$$\beta = \frac{\gamma_{\alpha\beta}}{\gamma_{\beta\alpha}}\left(1 - \frac{\lambda_{\alpha\beta}}{\lambda_\alpha}\frac{\gamma}{\gamma_{\alpha\beta}}\right).\tag{6.33}$$

Using the expressions for $\Gamma_{\alpha\beta}$ and $\Gamma_{\beta\alpha}$, this equation can be reduced to the following form, which contains only the density of states in addition to the usual constants λ_α and $\lambda_{\alpha\beta}$:

$$\beta = \frac{v_\beta}{v_\alpha}\left[1 - \frac{\lambda_{\alpha\beta}}{\lambda_\alpha}\left(\frac{v_\alpha}{v_\beta} + 1\right)\right].\tag{6.34}$$

In Eq. (6.32), T_c^0 is the critical temperature of an isolated α system $[(T_c^0 \approx \tilde{\Omega}\exp(-1/\lambda_\alpha)$; for simplicity, we put $\mu^* = 0)$. If we put $\lambda_{\alpha\beta} = 0$, we obtain the expression (6.18) describing the Cooper limit for the proximity effect. Note that $T_c^0 \ll \tilde{\Omega}$ (see above), and therefore the proximity channel leads to the inequality $T_c < T_c^0$. It is interesting that the mixed channel (second term in Eq. (6.33)) is favorable for superconductivity. We will discuss the results we obtain below (see Section 6.4.2.4).

6.4.2.2. Strong coupling

In this section, let us focus on the case when $2\pi T_c$ is much larger than the phonon energy, Ω: $T_c \gg \tilde{\Omega}/2\pi$. This corresponds to a large value of λ_α (cf. Section 2.3.3). In order to calculate T_c, one should eliminate the function Δ^β from the system of equations (6.23, 6.24). In addition, it is necessary to include the two-step processes in the renormalization function. As a result, we obtain

$$\Delta_n^\alpha Z_n^\alpha = \sum_{n'} (\lambda_\alpha D_{nn'} - \tilde{\mu}) \frac{\Delta_{n'}^\alpha}{|2n' + 1|} + \lambda_{\alpha\beta}\gamma_{\beta\alpha} \sum_{n'} D_{nn'}(|2n + 1|^{-1} + |2n' + 1|^{-1})$$

$$\times \frac{\Delta_{n'}^\alpha}{|2n + 1|} + \frac{\gamma_{\alpha\beta}\gamma_{\beta\alpha}}{|2n + 1| + \gamma_{\beta\alpha}} \frac{\Delta_n^\alpha}{|2n + 1|}, \tag{6.35}$$

$$Z_n^\alpha = 1 + \lambda_\alpha |2n + 1|^{-1} \sum_{n'} D_{nn'}(2n + 1)/|2n' + 1| + \frac{\gamma_{\alpha\beta}}{|2n + 1|}$$

$$+ \lambda_{\alpha\beta}\Gamma_{\beta\alpha}|2n + 1|^{-1} \sum_{n'} D_{nn'}(|2n + 1|^{-1} + |2n' + 1|^{-1}) \frac{2n' + 1}{|2n' + 1|}. \tag{6.36}$$

Here

$$D_{nn'} = \frac{v^2}{v^2 + (n - n')^2}.$$

The quantities v and γ_{ik} are defined by Eq. (6.28).

Introducing the function

$$\Delta_n = \Delta_n^\alpha \left(1 + \frac{\gamma_{\alpha\beta}}{|2n + 1| + \gamma_{\beta\alpha}} \right) \tag{6.37}$$

we obtain, after some manipulations,

$$\Delta_n = \sum_{n' \geq 0} \left[\lambda_\alpha \left(\frac{v^2}{v^2 + (n - n')^2} + \frac{v^2}{v^2 + (n + n' + 1)^2} \right) - 2\tilde{\mu} \right.$$

$$- \lambda_\alpha \delta_{nn'} \left[\sum_{n''} D_{nn''}(2n'' + 1)|2n'' + 1|^{-1} \right] \frac{\Delta_{n'}}{\left(1 + \dfrac{\gamma_{\alpha\beta}}{|2n' + 1| + \gamma_{\beta\alpha}} \right)|2n' + 1|}$$

$$+ \lambda_{\alpha\beta}\gamma_{\beta\alpha} \sum_{n' \geq 0} \left[\left(\frac{v^2}{v^2 + (n - n')^2} + \frac{v^2}{v^2 + (n + n' + 1)^2} \right) \right.$$

$$\times (|2n + 1|^{-1} + |2n' + 1|^{-1}) \Big]$$

$$- \left[\delta_{nn'} \sum_{n''} D_{nn''}(|2n + 1|^{-1} + |2n'' + 1|^{-1}) \frac{2n'' + 1}{|2n'' + 1|} \right]$$

$$\times \frac{\Delta_{n'}}{\left(1 + \dfrac{\gamma_{\alpha\beta}}{|2n' + 1| + \gamma_{\beta\alpha}} \right)|2n' + 1|}. \tag{6.38}$$

We neglect the term $\sim \lambda_{\alpha\beta}\lambda_{\beta\alpha}$, because this term contains a product of two-phonon Green's functions and, therefore, the additional small parameter $v^2 \ll 1$. We assume also that $\gamma_{\beta\alpha} \ll 1$.

It is convenient to use the matrix method developed in [25], see Section 2.3.3. Introducing the function

$$\varphi_n = \Delta_n(2n + 1)^{-1/2}\left[1 + \frac{\gamma_{\alpha\beta}}{|2n + 1| + \gamma_{\beta\alpha}}\right]^{-1/2} \tag{6.39}$$

we arrive at the following matrix equation:

$$\varphi_n = \sum_m R_{nm}\varphi_m. \tag{6.40}$$

Here

$$R_{nm} = (2n + 1)^{-1/2}(2m + 1)^{-1/2}\left[\frac{(2n + 1 + \gamma_{\beta\alpha})}{(2n + 1 + \gamma)}\frac{(2m + 1 + \gamma_{\beta\alpha})}{(2m + 1 + \gamma)}\right]^{-1/2}\tilde{F}_{n,m;v}, \tag{6.41}$$

$$\tilde{F}_{n,m;v} = A\{[v^2 + (n - m)^2]^{-1} + [v^2 + (n + m + 1)^2]^{-1}$$

$$- \delta_{nm}\sum_{l=0}^{2n}[v^2 + (n - l)^2]^{-1}\}$$

$$+ B\left\{\{[v^2 + (n - m)^2]^{-1} + [v^2 + (n + m + 1)^2]^{-1}\}\right.$$

$$\times (|2n + 1|^{-1} + |2m + 1|^{-1}) - \delta_{nm}\sum_{n''}[v^2 + (n - n'')^2]^{-1}(|2n + 1|^{-1}$$

$$+ |2n'' + 1|^{-1})\frac{2n'' + 1}{|2n'' + 1|}\right\} - 2\tilde{\mu}, \tag{6.42}$$

$$A = \lambda_\alpha v^2, \qquad B = \lambda_{\alpha\beta}\gamma_{\beta\alpha}v^2.$$

The matrix equation can be solved (see Section 2.3.3) and we arrive at the following expression for T_c:

$$T_c \simeq (2\pi)^{-1}(\Lambda_{\text{eff}})^{1/2}\tilde{\Omega}, \tag{6.43}$$

$$\Lambda_{\text{eff}} = \frac{\lambda_\alpha + 2\lambda_{\alpha\beta}\gamma_{\beta\alpha}}{1 + \gamma_{\alpha\beta}(1 + \gamma_{\beta\alpha})^{-1} + 2\tilde{\mu}}. \tag{6.44}$$

If we neglect the phonon-mediated interlayer channel ($\lambda_{\alpha\beta} = 0$), then T_c will be depressed by the proximity effect (factor $\sim \gamma_{\alpha\beta}$ in the denominator). On the other hand, the mixed channel, which combines the intrinsic proximity effect with the phonon-mediated process, enhances T_c. If we neglect both channels ($\lambda_{\alpha\beta} = \gamma_{\alpha\beta} = 0$), we obtain the equations (2.58) for $\mu = 0$ and (2.60) for $\mu \neq 0$. Thus, Eq. (6.43) is a generalization of (2.60) for the case of induced superconductivity. In addition, Eq. (6.43) represents the equation which allows us to determine T_c, because the right-hand side depends on T_c (remember that $\gamma_{\alpha\beta} = \Gamma_{\alpha\beta}/\pi T_c$).

6.4.2.3. *Intermediate coupling*

The intermediate coupling regime corresponds to λ_α of the order of 1 or $2\pi T_c$ of the same order of magnitude as the phonon energy. In this case, we cannot make any of the simplifying assumptions of the previous two cases but can nevertheless still use the matrix method, keeping all the terms in the system of equations with the pure two-band term. In this intermediate coupling case we must consider enough orders to guarantee convergence of the determinant, rather than using just the first order, which was sufficient for strong coupling. Consequently, there is no simple analytical expression but only a numerical solution. We checked the convergence by increasing the order of the matrix until the changes from one order to the next were less than 5% and then used the higher order.

To illustrate the power of the matrix method we have plotted $T_c/\tilde{\Omega}$ as a function of the ratio of densities of states between the α band and the β band for a selection of parameters. For Fig. 6.6, the value of λ_α was chosen

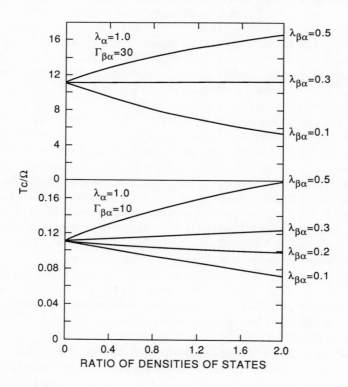

Fɪɢ. 6.6.

to be 10 K and several curves are plotted for various values of $\lambda_{\beta\alpha}$. In this figure the interplay of the various terms can readily be seen. For small values of $\lambda_{\beta\alpha}$ the pure proximity effect dominates and T_c is depressed as states are added to the β band. Conversely, for larger values of $\lambda_{\beta\alpha}$ the trend is reversed and ultimately the two-band and mixed term predominate. In the upper figure the value of $\Gamma_{\alpha\beta}$ has been increased to 30 K and again T_c/Ω is plotted as a function of the ratio of the densities of states for various values of $\lambda_{\beta\alpha}$. The trends are again clear except that a larger proximity term has a stronger effect and requires a larger value of $\lambda_{\beta\alpha}$ to compensate. For all the data μ was chosen to be zero for simplicity. As a cross check, the value of $T_c/\tilde{\Omega}$ at zero for the ratio of the densities of states should correspond to the single-band case. Our result in this case is fully consistent with the value obtained by using the expression (2.61) for T_c that is valid for any strength of the coupling.

6.4.2.4. *Discussion*

The equations (6.30), (6.32), and (6.43) describe the phenomenon of induced superconductivity. Although we have described several distinct cases, they all have some common features which we will discuss. As can be seen in the above expressions for T_c, the induced superconductivity has several channels. One of them is the intrinsic proximity effect which contains only $\gamma_{\beta\alpha}$. The intrinsic proximity effect induces the superconducting state for the β system, but at the same time depresses T_c. Another phonon-mediated channel, which contains only phonon matrix elements, also induced the superconducting state but has the opposite effect on T_c (enhances T_c). This effect is due to additional phase space for the phonon subsystem.

A very important feature of our theory is the appearance of the mixed channel. This channel contains the product of $\gamma_{\beta\alpha}$ and $\lambda_{\alpha\beta}$ and corresponds qualitatively to the following dynamics: Two carriers from the α group form a Cooper pair in the β band because of the phonon exchange ($\lambda_{\beta\alpha}$) and then this pair tunnels back to the α band as in the usual proximity effect.

The mixed channel is also important for induced superconductivity and is favorable for T_c. This means that when the proximity effect is combined with a phonon-mediated channel it is no longer just a negative influence on T_c. This is of course opposite to our conventional view of the proximity effect, in which it always has a negative influence on T_c.

Thus, for the critical temperature we have two opposite trends; one of them is due to the pure proximity channel which always depressed T_c, and the opposite trend is due to the pure phonon-mediated channel and to the mixed channel. This competition is an important factor for various real systems including the cuprates (see Chapter 7).

On the other hand, the picture for the induced energy gap is entirely different. In this case, we have two contributions from the phonon-mediated channel and the proximity channel and both of them produce a positive contribution to the induced energy gap (see below, Section 6.4.3). That is why, if we diminish the effect of induced superconductivity, that is, we decrease the values of $\lambda_{\beta\alpha}$ and $\gamma_{\beta\alpha}$, for example, by increasing the layer separation, then the induced energy gap is drastically depressed, whereas T_c changes much more slowly.

Our model is a generalization of the usual proximity effect and two-band picture. It can be seen directly from Eqs. (6.31, 6.33) that if we neglect the proximity term, for example, we end up with the usual two-gap model, whereas if we neglect the phonon-mediated contribution ($\lambda_{\beta\alpha} = 0$) we obtain an expression describing the proximity effect. Therefore our model contains all these limiting cases, but it cannot be treated as the sum of the contributions since it contains the very important mixed term.

6.4.3. Spectroscopy

6.4.3.1. Induced energy gap

This section is concerned with the spectroscopy of the system, containing an intrinsically superconducting (α) and intrinsically normal (β) subsystem with a charge transfer between them. Let us evaluate the induced energy gap $\varepsilon_\beta(0)$ at $T = 0$ K.

Each subsystem is characterized by its own density of states, so that

$$N_i(\omega) = \mathrm{Re}\left\{\frac{\omega}{[\omega^2 - \Delta_i^2(\omega)]^{1/2}}\right\}, \qquad i \equiv \{\alpha, \beta\}. \tag{6.45}$$

Here $\Delta_\alpha(\omega), \Delta_\beta(\omega)$ are the order parameters which are the analytic continuations of the thermodynamic order parameters $\Delta_\alpha(\omega_n)$ and $\Delta_\beta(\omega_n)$ at $T = 0$ K.

The appearance of an induced order parameter $\Delta_\beta(\omega)$ leads to the induced energy gap, ε_β, which is close to the value of a frequency corresponding to a peak in the density of states $N_\beta(\omega)$. The function N_α also has a peak at $\omega \approx \varepsilon_\beta$ (it is shown in [14] for the usual proximity system; the system has a single energy gap). Nevertheless, N_α has a strong peak at $\omega \approx \varepsilon_\alpha$ which is close to the value of the energy gap for an isolated α subsystem. As was noted above, the phenomenon of two-gap superconductivity should be understood as the presence of two peaks in the superconducting density of states. Any spectroscopy of the material will display this structure.

The energy gap ε_β can be obtained from the equations similar to Eqs. (6.23, 6.24), taken at $T = 0$ K (one should make an analytical continuation). One should make the replacements

$$\pi T \sum_n \rightarrow \int d\omega, \qquad F_i(\omega_n) \rightarrow F_i(\omega) = \Delta_i(\omega)[\omega^2 + \Delta_i^2(\omega)]^{-1/2}.$$

Another approach is based on the direct use of the equations for the usual time-dependent Green's function:

$$\Delta_\alpha(\omega) = Z_\alpha^{-1}(\omega) \int d\omega' \, (\lambda_\alpha D_{\omega,\omega'} - \tilde{\mu}) F_\alpha^+(\omega') + \Gamma_{\alpha\beta} Z_\alpha^{-1}(\omega) F_\beta^+(\omega), \quad (6.46)$$

$$\Delta_\beta(\omega) = Z_\beta^{-1} \int d\omega' \, \lambda_{\beta\alpha} D_{\omega,\omega'} F_\alpha^+(\omega') + \Gamma_{\beta\alpha} Z_\beta^{-1} F_\alpha^+(\omega). \quad (6.47)$$

Here

$$D_{\omega,\omega'} = \frac{\Omega}{2} [(\omega - \omega' - \Omega + i\delta)^{-1} + (\omega - \omega' + \Omega - i\delta)^{-1}]$$

is a phonon Green's function, $F_i(\omega) = \Delta_i(\omega)[\omega^2 - \Delta^2(\omega)]^{-1}$. We will not write out explicit expressions for the renormalization functions $Z_\alpha(\omega)$, $Z_\beta(\omega)$ (see Section 2.2).

Let us evaluate the energy gap ε_β. Assume that $\Gamma \ll \varepsilon_\alpha$ and $\lambda_{\alpha\beta} \ll \lambda_\alpha$; this means that the pairing in the subsystem α is mainly determined by its intrinsic attraction (λ_α). Assume also that $\lambda_{\beta\alpha} \ll 1$. The energy gap is a root of the equation $\omega = \Delta_\beta(\omega)$.

The energy gap is determined by the relation

$$\varepsilon_\beta \approx \frac{\lambda_{\beta\alpha}}{\lambda_\alpha} (1 + \lambda_\alpha)\Delta_\alpha(0) + \Gamma_{\beta\alpha}. \quad (6.48)$$

Indeed, if we are interested in the region $\omega \approx \varepsilon_\beta \ll \varepsilon_\alpha$, then $\Delta_\alpha(\omega) \approx \Delta_\alpha(0)$. Moreover, $Z_\beta(\omega \approx \varepsilon_\beta) \approx 1$, since $\Gamma_{\beta\alpha}[\Delta^2(0) - \varepsilon_\beta]^{1/2} \approx \Gamma_{\beta\alpha}/\Delta_\alpha(0) \ll 1$. In addition, the first term on the right-hand side of Eq. (6.46) can be written in the form

$$\lambda_{\beta\alpha} \int d\omega' \, D_{\omega,\omega'} F^{\alpha+}(\omega')_{|\omega \approx \varepsilon_\beta} \approx \frac{\lambda_{\beta\alpha}}{\lambda_\alpha} (1 + \lambda_\alpha)\Delta_\alpha(0).$$

One should stress a very important feature of the charge transfer. The pure proximity effect depresses T_c (see Eqs. (6.31, 6.32, 6.44)). The situation for the induced energy gap is entirely different. One can see directly from Eq. (6.48) that both phonon-mediated and proximity channels make positive contributions to the formation of the induced gap ε_β.

If the coupling is weak ($\lambda_\alpha \ll 1$),

$$\varepsilon_\beta \approx \frac{\lambda_{\beta\alpha}}{\lambda_\alpha} \varepsilon_\alpha + \Gamma_{\beta\alpha}. \quad (6.49)$$

Here ε_α is an energy gap for the α subsystem. For the strong coupling case, $\Delta_\alpha(0) \neq \varepsilon_\alpha$, namely, $\Delta_\alpha(0) < \varepsilon_\alpha$ (Section 2.4). If $\lambda_{\beta\alpha}$ is small ($\lambda_{\beta\alpha} \ll \lambda_\alpha$), then

$\varepsilon_\beta \approx \Gamma_{\beta\alpha}$. Therefore, the induced energy gap is determined mainly by the proximity channel.

Note also that each of the subsystems is characterized not only by its own order parameter and energy gap but also by its coherence length $\xi_i = \hbar V_F^i/\varepsilon_i$ ($i \equiv \alpha, \beta$); the values of ξ_α and ξ_β may differ in a very noticeable way.

6.4.4. Magnetic impurities. Gapless induced superconductivity

Consider an interesting case when the β system contains magnetic impurities. In other words, the Hamiltonian H_0 contains an additional term with an interaction $V(r) = v_1(r) + v_2(r) (S\sigma)$. The induced energy gap ε_β is strongly affected by the presence of the impurities, whereas the change in T_c is relatively small. Such behavior of a system with induced superconductivity is very different from that for the usual, one-gap superconductor [26], when both the gap and T_c undergo a drastic decrease.

At some critical value of the concentration (see below) the induced gap, ε_β, becomes equal to zero (although the order parameter $\Delta_\beta \neq 0$). Nevertheless, the density of states $N_\alpha(\omega)$ continues to have a strong peak at $\omega \approx \varepsilon_\alpha$; ε_α is an intrinsic gap for the α subsystem. As a result, we are dealing with a phenomenon of an induced gapless superconductivity.

Consider this case in more detail. The order parameter at $T = 0$ K is described by Eq. (6.47). We are using a method similar to [27] developed for the usual proximity system. In the presence of the magnetic impurities the renormalization function, Z_β, contains an additional term which, after the averaging over impurity positions [26] can be written in the form

$$Z_\beta^m = \Gamma_M^{(1)}[\Delta_\beta^2(\omega) - \omega^2]^{-1/2},$$

where

$$\Gamma^{(1)} = n_i \pi v[|v_1|^2 + \tfrac{1}{2}J^2 S(S + 1)|v_2|^2]$$

(n_i is the concentration of the impurities). The equation for the self-energy part Σ_2 also contains an additional term.

At some value of the parameter $\Gamma_M = \Gamma_M^c$ the induced energy gap ε_β disappears. The value Γ_M^c is characterized by the appearance of a nonzero value of the density of states N_β (see Eq. (6.45)) at $\omega > 0$ and is equal to

$$\Gamma_M^c = [(\Delta_\beta^{ph} + \Gamma_{\alpha\beta})\Delta_\alpha^{ph} + \Delta_\beta^{ph}\Gamma_{\alpha\beta}](\Delta_\alpha^{ph} + \Gamma_{\alpha\beta})^{-1}. \qquad (6.50)$$

If $\Gamma_{\alpha\beta} \ll \varepsilon_\alpha$, we have

$$\Gamma_M^c \approx \varepsilon_\beta^0 \approx \frac{\lambda_{\beta\alpha}}{\lambda_\alpha}(1 + \lambda_\alpha)\varepsilon_\alpha + \Gamma_{\beta\alpha}. \qquad (6.51)$$

If, for example, $\lambda_{\beta\alpha} = 0.25$, $\lambda_\alpha = 2.5$, $\varepsilon_\alpha = 20\,\text{meV}$, $\Gamma_{\beta\alpha} = 5\,\text{meV}$, then $\Gamma_M^C = 12\,\text{meV}$.

In the opposite limit, which corresponds to weak coupling ($\lambda_\alpha \ll 1$) and a relatively large tunneling amplitude, so that $\Gamma_{\alpha\beta} \gg \varepsilon_\alpha$, one obtains

$$\Gamma_M^C = \left(\frac{v_\alpha}{v_\beta} + \frac{\lambda_{\beta\alpha}}{\lambda_\alpha}\right)\varepsilon_\alpha. \qquad (6.52)$$

The critical concentration can be estimated from the relation

$$l_M^C \approx v_F^\beta \tau_M^C \approx \hbar v_F^\beta (\Gamma_M^C)^{-1} \approx \xi_\beta. \qquad (6.53)$$

Therefore, the presence of a small amount of magnetic impurities leads to the gapless state ($\varepsilon_\beta = 0$). However, it does not lead to a noticeable change in the critical temperature (the shift $\Delta T_c \ll T_c$). This is a very important property of the model which makes the system with induced superconductivity entirely different from the usual gapless superconductivity [26]. Qualitatively, this feature is due to the fact that T_c of the system is determined, mainly, by the state of the intrinsically superconducting α subsystem.

Thus, the presence of magnetic impurities drastically affects the value of the induced energy gap without any noticeable impact on the critical temperature.

6.4.5. Major parameters

The superconductivity state of the system is characterized by three parameters: λ_α, $\lambda_{\beta\alpha}$, and $\gamma_{\beta\alpha}$ (note that $\lambda_{\alpha\beta}(\gamma_{\alpha\beta}) = \lambda_{\beta\alpha}(\gamma_{\beta\alpha})r$, $r = v_\beta/v_\alpha$ is the ratio of the densities of states; for simplicity we do not consider μ). Each of these parameters has a clear meaning. A description of many properties is more complicated than in usual BCS theory which contains one parameter λ. Because of the one-to-one correspondence between T_c and λ, all electromagnetic, transport, and other properties can be expressed in BCS in terms of one measured parameter, namely T_c. Such an important quantity as the energy gap $\varepsilon(0)$ is related to T_c in a simple way: $\varepsilon(0) = 1.76T_c$.

Properties of the α–β system with an induced superconductivity cannot be expressed in terms of T_c only, because of the existence of three parameters, not one. As a result, there is no universal relation between $\varepsilon_i(0)$ and T_c. Various properties of the system can be described in terms of $\varepsilon_\alpha(0)$, $\varepsilon_\beta(0)$, and T_c; these are three experimentally measured quantities. However, if we want to calculate T_c we should determine the values of the basic parameters λ_α, $\lambda_{\beta\alpha}$, $\gamma_{\beta\alpha}$ (see Section 7.4.1).

6.4.6. Conventional superconductors

If we study an N–S–N sandwich (or the corresponding multilayer structure) and we are interested in the dependence of T_c on L_N, where L_N is the thickness of the normal layer, then one can expect that for small L_N, T_c increases with increasing L_N, because the phonon-mediated and "mixed" channels are dominant over the pure proximity channel. A further increase in L_N leads to the proximity effect becoming dominant and depressing T_c. Such an effect probably contributes to the initial increase in T_c discussed in Section 6.2, along with the change in the phonon spectrum. The manifestation of the phonon-mediated channel can be observed directly if the S and N films have a similar phonon spectra.

An induced gap has been observed [28] in conventional S–N–S sandwich structures consisting of lead–bismuth superconducting films surrounding a thin silver layer. From measurements of the temperature dependence of the thermal conductivity in the superconducting and normal states it was found that for the thinnest silver layers the induced gap appeared to saturate at a value of 0.375 meV. These experiments are quite consistent with the model presented above.

It would be interested to study tunneling into an S_α–N_β sandwich by using the systems S_α–N_β–I–S′ and N_β–S_α–I–S′ (I is an insulator and S′ is a counterelectrode). Such experiments allow the determination of the intrinsic and induced energy gaps ε_α and ε_β.

7

HIGH T_c CUPRATES

7.1. Introduction

In this chapter we are going to focus on a description of the high-transition-temperature cuprate superconductors. Their discovery stimulated much of the theoretical development that is described in the previous chapters. At the present time there are a large number of compounds in this family but the original discovery by Bednorz and Mueller [1] was a LaBaCuO compound with a transition temperature of about 30 K. These compounds were studied earlier in detail by Raveau and his coworkers [2b] who demonstrated that they were metallic. A breakthrough with a transition temperature above liquid nitrogen temperature was achieved in the compound $YBa_2Cu_3O_{7-x}$ [3]. Since then the number of unique materials has risen to over 50. Some compounds have transition temperatures in excess of 100 K: BiSrCaCuO [4] including SrCaCuO [4b] and TlBaCaCuO [5]. All cuprates have one feature in common, namely CuO planes which form a quasi-2D layered compound. The CuO plane is a key unit essential for high T_c.

During the last several years there has been significant progress in understanding the properties of the high T_c oxides (see references given in Chapter 1). The situation now is entirely different from that initial time right after the discovery of high T_c superconducting oxides when the nature of the new materials was completely unknown. Of course, there are still many unresolved problems (this is also true with respect to conventional superconductivity).

In this chapter we describe our current understanding of the nature of the normal and superconducting states of these unusual materials. It is based on several papers we have published over the last few years [6–16].

Initially we describe the normal properties of the cuprates. Superconducting properties which do not depend directly on the mechanism of high T_c (behavior in an a.c. field, peculiar two gap superconductivity, positron annihilation, etc.) will be discussed afterward. Finally, we will concentrate on the origin of high T_c.

7.2. Normal properties

We will focus on the doped or metallic state. Only the doped or metallic materials undergo transitions into the superconducting state; as a result, our analysis is directly related to the origin of high T_c.

The LaSrCuO system will be given special attention because of the simplicity of its structure. At the same time LaSrCuO contains the major structural unit, namely the CuO plane (this can also be said of single TlSrCuO layered compound, or the infinite layered CaSrCuO). This system plays a role similar to the hydrogen atom in atomic physics; it is the best test system for understanding the basic principles of high-temperature superconductivity.

7.2.1. One-particle excitations

Let us evaluate the major parameters such as the effective mass m^*, the Fermi energy E_F, and the Fermi momentum p_F. In other words, we are talking about the usual parameters whose values are well known for usual metals (see, e.g., [17]). The major parameters for LaSrCuO are presented in Table 7.1 (we will discuss the properties of the YBaCuO below). The values of these parameters have been evaluated by us in [6a,b]. The method used in the evaluation will be described below, but initially we want to stress the unique features of the material. In order to do this, we have also included typical parameters for conventional metals.

One can see directly from Table 7.1, that the cuprates are characterized by uniquely small values of the Fermi energy, E_F and the Fermi velocity, v_F. The Fermi energy is almost two orders of magnitude smaller in LaSrCuO than in usual metals.

We believe that the small values of the Fermi energy and the Fermi velocity along with large anisotropy are the important features of the cuprates.

Table 7.1

Quantity	Conventional metals	$La_{1.8}Sr_{0.2}CuO_4$
m^*	$1-15m_e$	$5m_e$
k_F (cm^{-1})	10^8	3.5×10^7
v_F (cm^{-1}/s)	$(1-2) \times 10^8$	8×10^6
E_F (eV)	$5-10$	0.1

We would like to point out that the effective mass $m^* = m(0)$ appears to be relatively large ($m^*(0)$ corresponds to $T = 0$ K). It is important to emphasize that this value of the effective mass represents a so-called renormalized value, and it can be an experimentally observed quantity. It enters the density of states and can be determined, for example, by heat capacity or de Haas–van Alphen measurements. It is not equal to the band value of the effective mass m^b (that is to say to the value for the "frozen" lattice) obtained from electronic structure calculations or from magnetic susceptibility measurements. There is a simple relation (see Section 2.5) $m^*(0) = m^b(1 + \lambda)$, where λ is the electron–phonon coupling constant.

On the other hand, the Fermi momentum, p_F, is comparable with values found in conventional superconductors. This is important, because it means that there is sufficient phase space for pairing. A large value for the Fermi momentum does not contradict the small value of carrier concentration because of the quasi-2D nature of the Fermi surface.

At this point we will describe our approach [6a,b]. We employ the method which was widely used during the "Golden Age" of physics of metals, namely "Fermiology" (see, e.g., [17, 18]). In doing this, we assume that the superconducting high T_c oxides are metallic systems, or, in other words, they are characterized by the presence of a Fermi surface (indeed, by definition, the metallic state is a solid with a Fermi surface). Anisotropy of the system is reflected in the shape of the Fermi surface, its deviation from a spherical shape. As in a typical metal, the parameters of the Fermi surface can be reconstructed using appropriate experimental data.

Of course, the use of the Fermi liquid concept ([19], see also [20]) for unknown materials like the cuprates is not obvious. There were many doubts related to this issue. Our conviction about the applicability of the concept of the Fermi surface to the doped cuprates has been based on our analysis which resulted in the set of parameters in Table 7.1. As mentioned before, the Fermi momentum appears to be relatively large, and thus it allows us to define the Fermi surface.

Moreover, the measurements of heat capacity [21] have displayed the linear law $C_{el} = \gamma T$. It is important that the Sommerfeld constant γ strongly depends on magnetic field: $\gamma = \gamma(H)$. This is related to the contribution of the vortex region which can be treated as a normal; the dependence of $C_{el} \propto T$ can be interpreted as a direct indication of the presence of the Fermi surface since the well-known derivation of this dependence is based on the presence of the Fermi surface.

Recent experimental data (see below, Section 7.6) confirmed the viewpoint [6, 7] about the presence of the Fermi surface. Among these data are photoemission spectroscopy, positron annihilation, and de Haas–van Alphen measurements.

In order to evaluate the major parameters (see Table 7.1) we assumed the Fermi surface for LaSrCuO to be cylindrically shaped (this shape corresponds to a layered structure). Of course the interlayer transitions lead to small deviations from the cylindrical shape. Such deviations are not important for our present purpose, namely for the estimation of values of the major parameters. Not also, that we are not assuming that the Fermi curve is a circle (the Fermi curve is defined as the cross-section of the Fermi surface by the plane p_z = constant). This approach is applicable to hole as well as electron carrier materials. For example, the hole surfaces at the corners of the first zone can be viewed in the quasi-2D case as a cylinder as can be seen from a simple translation in momentum space. As is well known, Fermiology as a conventional technique in the physics of metals (see, e.g. [18]) is based on reconstruction of the parameters of the Fermi surface with use of some special measurements. Cyclotron resonance data, de Haas–van Alphen, or Shubnikov–de Haas measurements are examples of the usual methods employed in Fermiology. It is important to stress that for layered conductors one can also use heat capacity data. Indeed the Sommerfeld constant for such metals is proportional to m^*/d_c, where m^* is the cyclotron mass and d_c is the interlayer distance; thus measurements of heat capacity provide information similar to that from cyclotron resonance experiments.

Let us start with the expression for the total energy,

$$E = 2 \int \frac{d\mathbf{k}\, dp_x\, \varepsilon f}{(2\pi \hbar)^3}, \tag{7.1}$$

where f is the Fermi function, integration over p_z is restricted by $|p_{z\,max}| = \pi/d_c$ where d_c is the interlayer distance. With the use of Eq. (7.1) and making use of the transformation into integrals over constant energy curves and energy, we obtain the following expression for the Sommerfeld constant $\gamma = C_e/T$:

$$\gamma = (\pi/3h^2)m^* k_B^2 d_c^{-1}, \tag{7.2}$$

where the average effective mass is defined as

$$m^* = (2\pi)^{-1} \int dl\, v_\perp^{-1}, \tag{7.3}$$

and the integration is taken over the Fermi curve $E(k) = E_F$, p_z = constant, $v_\perp = (\delta \varepsilon/\delta k)_F$. In this analysis we have assumed that the effective mass as defined in Eq. (7.3) does not strongly depend on energy. Note that this mass corresponds to the cyclotron mass, and in the case of a simple parabolic band this cyclotron mass is equal to the usual effective mass.

Equation (7.2) can be used to determine the value of m^*:

$$m^* = 3h^2 d_c \gamma/\pi k_B^2. \tag{7.4}$$

It is important to stress that both Eq. (7.4) as well as Eq. (7.5) (see below) are valid for a cylindrical Fermi surface. This symmetry reflects the layered structure of the material. It is interesting to note that the density of states v, and, therefore, the Sommerfeld constant ($\gamma \propto v$), for an isotropic 3D system is proportional to $m^* p_F$. As a result, v_{3D} increases with carrier concentration, n, since $p_F \propto n^{1/3}$. That is why in the BCS theory a small carrier concentration is considered a negative factor (according to BCS theory T_c depends directly on the density of states); as a consequence, we observe low values of T_c for semiconducting superconductors. The situation is different for layered metals. Then the density of states $v_L \propto m^* d_c^{-1}$ does not depend directly on the carrier concentration. As a result, the relatively small carrier concentration in the cuprates does not contradict the large value of T_c in these new materials.

Equation (7.4) expresses m^* in terms of the experimentally measured quantity γ. In addition, one can derive an expression for the Fermi energy which has the form for a cylindrical Fermi surface:

$$E_F = \pi^2 k_B^2 n / 3\gamma. \tag{7.5}$$

Equation (7.5) expresses the value of the Fermi energy in terms of experimentally measured parameters: the carrier concentration n and the Sommerfeld constant γ. The values of $\gamma \equiv \gamma(0)$, the zero temperature limit, and n can be obtained from heat capacity data (see [21]) and Hall effect data (see, e.g., [22]). Note that the determination of γ is not a simple task because the system is in the superconducting state and the critical field is large. This problem has been solved in [21] by analyzing the dependence of γ on magnetic field, H, using a model appropriate for a type II superconductor. The authors [21] observed a linear dependence: $\gamma \propto H$ in the region near $T = 0$ K. As was noted above, the dependence γ upon H is caused by the contribution of the vortex region. At $H = H_{c2}$ the material becomes a normal metal, and, therefore $\gamma(H_{c2}) = \gamma(0) = \gamma_n(0)$, with use of a linear approximation for $\gamma(H)$ (this is analogous to the well-known Bardeen–Stephen law for flow resistance), one can obtain $\gamma \approx H_{c2}(\delta\gamma/\delta H)$. The linear approximation is valid up to values of H approaching H_{c2}; it fails only in the small region near H_{c2} (because of the vortex–vortex interaction) and at very low fields near the Meissner region. Therefore, the approximation works very well for an estimation of $\gamma(0)$. The value of $H_{c2} \approx 90$ T can be used in order to evaluate $\gamma(0)$ for the LaSrCuO compound. Note that the value of γ and other parameters we obtained can be used in order to calculate H_{c2} (see Section 7.3.1). We obtained the value close to 90 T and it demonstrates a self-consistency of our approach.

Furthermore, the carrier concentration for the LaSrCuO compound can be determined from Hall effect data and is approximately equal to 3×10^{21} cm^{-3} [22] (a similar value has been obtained using a chemical method [23]). We want to point out that the ability to use the Hall effect is related to the simple band structure of this compound, which leads to a weak temperature dependence of the Hall coefficient. The situation with the YBaCuO compound is more complicated, because this system contains two sets of carriers (two different bands, see below, Section 7.3) and this leads to a strong temperature dependence of the Hall coefficient.

Using the expressions (7.4) and (7.5) one can evaluate the major parameters presented in Table 7.1. Let us stress again the small values of E_F and v_F, which are much smaller than in conventional metals.

7.2.2. Collective excitations

7.2.2.1. Phonons

We think that phonons play a key role in high T_c superconductivity (see below, Section 7.4.5). Such statements may sound old-fashioned, because it means a similarity with conventional superconductors. However, we stress that we are dealing with an exotic phonon system in the cuprates. Cuprates are polyatomic systems (they contain four or more components), and their chemistry and structure are more complicated than those of ordinary superconductors and their phonon spectrum is very rich. This can be seen directly from the neutron spectroscopy data (see, e.g., [24, 25]). In addition, we indicate some unique (relative to conventional superconductors) features of the lattice dynamics of cuprates. The distinctive features are: (1) Presence of low frequency optical modes; for example, the LaSrCuO compound has two modes at ~ 10 meV and ~ 20 meV; for YBaCuO $\Omega_{opt} \sim 25$ meV. Optical phonons have a large density of states, and this feature along with low frequency is favorable for the pairing (see (2.39)). (2) Anharmonicity. (3) Presence of the phonon-like acoustic plasmon mode (electronic "sound," see next section). As a result, the lattice dynamics is very peculiar and differs drastically from that in conventional metals.

Ferroelectricity of the perovskites is a definite manifestation of unusual lattice dynamics. This is why the initial motivation of Bednorz and Muller was quite logical.

It was interesting to find that in the doped conducting LaSrCuO compound neutron scattering [26] showed the tetragonal–orthorhombic transition to be driven by a zone boundary soft-tilt mode of the CuO units. A parallel may be drawn between these rigid-body-like motions of subunits of the high T_c cuprates and organic superconductors, in which the very nature of molecular structure leads to new orientational lattice modes—librons— which hybridize with the center-of-mass displacement longitudinal and transverse phonons [27].

7.2.2.2. Plasmons in layered cuprates. "Electronic" sound

The concept of plasmons is not a new one in solid state physics. Plasmons describe the collective motion of the carriers relative to the lattice. Usual metals are characterized by a dispersion relation $\omega = \omega_0 + aq^2$ for the plasmons; therefore, the plasmon spectrum has a finite value ω_0 at $q = 0$. In addition the value of ω_0 in usual metals is large (≈ 5–10 eV). The situation in layered materials is quite different. In fact, in the early 1970s there was a surge of interest in layered metals such as TaX (X = S, Se, Te), and intercalated graphite, which became superconducting. In contrast to isotropic metals, these materials have considerable anisotropy in their conductivities along and perpendicular to the layer planes. As a consequence, the screening properties of such a system are quite different from the isotropic case and were analyzed in some detail by various theorists (see, e.g., [28]). The main attention has been paid to the region of large frequencies $\omega > \kappa v_F$, where κ is the in-plane momentum).

The term "layered electron gas" (LEG) describes the basic physical concept, namely, an infinite set of two-dimensional layers of carriers, described as a 2D electron gas, separated by electronically inactive (insulating) spaced layers. It was shown for a single layer (see Section 4.6.2) that the collective (plasmon) mode frequency goes to zero for small in-plane wave vector κ: $\omega \sim \sqrt{\kappa}$. In the LEG model the Coulomb interaction between charge carriers in different layers is included exactly, while the polarization of the system is treated by including the electron–hole pair response calculated for a single layer with 2D plane-wave wave functions for the carriers. In addition, the dielectric response of the intervening, nonmetallic, region is included by using the low-frequency limit of its dielectric constant $\varepsilon \cong 3$–6 to screen the interaction between conducting layers. The entire spectrum of the LEG has been analyzed analytically in [9f].

A direct consequence of the layering of the carriers is that the spatial periodicity along the normal to the planes allows the definition of a corresponding wave vector $q_z = \pi/nd_c$ ($n = 1, 2, \ldots$), which labels the phase relations between charge oscillations on different planes. The calculation utilizes a 2D Fourier transform describing the in-plane Coulomb interaction, while the direction perpendicular to the layers is described by a discrete index. The fully screened interaction can be studied algebraically in the random-phase approximation (RPA) by calculating the polarizability in a single layer and solving the Dyson equation for the screened interaction. The layer plasmon dispersion relations is determined by the equation (in RPA):

$$1 - \Pi \Phi_0(\kappa, q_z) = 0, \tag{7.6}$$

where

$$\Phi_0 = \frac{2\pi e^2}{\varepsilon_m \kappa} \frac{\sinh(\kappa d_c)}{\cosh(\kappa d_c) - \cos q_z d_c} \tag{7.7}$$

(ε_m is the background dielectric constant, κ is in-plane momentum, and Π is a 2D polarization operator). The function $\chi = \operatorname{Re} \Pi$ has been evaluated in

[9c] and has the form (for $\omega > \kappa v_F$):

$$\chi = v[\omega(\omega^2 - v_F^2 \kappa^2)^{-1/2} - 1], \tag{7.8}$$

where v is a 2D density of states. As a result we arrive at the following expression for the plasmon frequency [9f]:

$$\omega = v_F \kappa \left(1 + \frac{\alpha^2}{4(\alpha + 1)}\right), \tag{7.9}$$

where

$$\alpha = (2m/\pi)\Phi_0. \tag{7.10}$$

Φ_0 is defined by Eq. (7.7). If $\alpha \gg 1$ we obtain

$$\omega = v_F \kappa \left(1 + \frac{\alpha}{4}\right). \tag{7.11}$$

This result (7.11) has been obtained in [28b] with the use of the hydrodynamic approximation in the region of small κ.

Equation (7.9) describes the structure of the plasmon band in layered conductors, such as cuprates (Fig. 7.1). Most strikingly, the completely in-phase motion of all layers $q_z = 0$ has a frequency ω_P, which corresponds closely to the isotropic (bulk) plasma frequency. All other branches with $q_z \neq 0$ go to zero frequency as their transverse wave vector κ (which is proportional to the inverse of the wavelength of the density oscillation in the plane) goes to zero. This breaking of the degeneracy of the standard three-dimensional plasmon excitation into different branches of charge oscillation modes filling in the low-frequency and transverse wave vector space of the spectrum is, in our view, a very important aspect of layered conductors; it is particularly important that the copper oxide high T_c superconductors belong to this category.

It is also possible to include a small hopping term t in the c-axis direction; one now finds a small gap at $\kappa = 0$ of order $2t$.

There are several novel consequences from the splitting of the collective excitation spectrum into layer plasmon branches. These arise specifically because of the existence of a low-frequency electronic response. The first of these is the possibility of pairing the carriers in the copper oxides by exchange of acoustic plasmons; we proposed such a contribution to the high superconducting transition temperatures in the layered cuprates [9a,b,c]. Our view is that such pairing occurs in addition to strong electron–phonon coupling (Section 7.5); it could also be treated as an effective decrease in μ^*.

A second consequence of the low-frequency plasmon spectrum is the appearance of a temperature-dependent term in the electron loss function

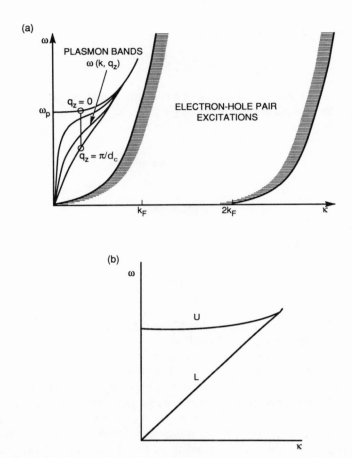

FIG. 7.1. Plasmons in a layered conductor: (a) plasmon band; (b) U and L branches.

because of the thermal population of the acoustic plasmon branches. Electron energy loss spectroscopy is a conventional technique for studying plasmon excitation in solids. For the ordinary 3D metals $\omega_{pl} \gg T$. As a result, the amount of total losses is independent of temperature. The picture is entirely different in the LEG. The passage of a charged particle through a layered conductor has been studied in [9g]. Calculation of the energy loss

$$\frac{\partial E}{\partial t} = (2\pi)^{-3} \int (\varepsilon_{\mathbf{p}} - \varepsilon_{\mathbf{p}-\mathbf{q}}) W_{\mathbf{q}} \, d^3\mathbf{q} \tag{7.12}$$

($\mathbf{q}$ is a momentum transfer and $W_{\mathbf{q}}$ is a total probability) has resulted in temperature-dependent picture. The observation of such a temperature-dependent picture is an interesting experimental problem. This dependence is a direct manifestation of a peculiar plasmon spectrum in the cuprates.

Finally, it is also possible to relate the layer plasmon spectrum to $\text{Im}(\varepsilon^{-1})$ where $\varepsilon = \varepsilon_1 + i\varepsilon_2$ is the dielectric function of the layered metal, with ε_1 and ε_2 the real and imaginary parts, respectively. One can obtain the dependence: $\text{Im}(\varepsilon^{-1}) \propto \tilde{\omega}[1 - \tilde{\omega}^2]^{-2}$, $\tilde{\omega} = \omega/\omega_v$ [10b].

Therefore, the plasmon spectrum in layered conductors such as cuprates represents not a single branch, but the plasmon band. The density of states is peaked at the upper ($q_z = 0$) and lower ($q_z = \pi/d_c$) boundaries. Qualitatively, one visualizes the plasmon band as a set of two branches (Fig. 7.1b). The upper (U) branch is similar to that in usual metals. A very important feature of the layered metals is the appearance of the lower (L) branch, which has an acoustic dispersion law and can be called "electronic" sound. The dispersion relation has the form (see Eqs (7.7), (7.9), (7.10)):

$$\omega_L = v_F \kappa [1 + \alpha_L^2(\alpha_L + 1)^{-1}]$$

$$\alpha_L = \frac{4me^2}{\varepsilon_m \kappa} \frac{\sinh(\kappa d_c)}{\cosh(\kappa d_c) + 1}$$

One should stress also that the slope of this acoustic branch is of order of v_F (not the sound velocity as for usual phonons); however, the value of v_F is small (see Table 7.1), and therefore, we have an additional phonon-like branch with relatively large phase space.

7.3. Superconducting properties

In this section we describe some unique features of the cuprates in the superconducting state. These particular features do not depend directly on the mechanism of the pairing in the oxides, that is, on the nature of the binding force. They are mainly due to the exotic normal properties discussed in the previous section.

7.3.1. Coherence length

The coherence length is one of the key parameters describing the superconducting state. Its value can be estimated from the well-known expression $\xi_0 = hv_F/2\pi T_c$. Using the values of the Fermi velocity (see Table 7.1), $v_F = 8 \times 10^6$ cm^{-1} and $T_c \approx 40$ K, we obtain $\xi_0 \approx 25$ Å. One can use a more precise definition of $\xi_0 = hv_F/\pi\varepsilon(0)$, where $\varepsilon(0)$ is the energy gap which is directly related to T_c, $\varepsilon(0) = aT_c$. In the BCS theory $a = 1.76$, but strong-coupling effects leads to an increase in a; e.g., for LaSrCuO $a = 2.5$ (see below, Section 7.4). As a result, a better estimate of ξ_0 is 20 Å. Note that estimates made by magnetic measurements are consistent with this calculation.

This value of the coherence length is very small relative to its value in the conventional superconductors ($\approx 10^3$–10^4 Å); this is a very important property of the new materials. The short coherence length in LaSrCuO is mainly due to a small value of the Fermi velocity (see Table 7.1), although the higher T_c also contributes to the decrease in ξ_0 ($\xi_0 \propto v_F/T_c$).

Using this value of the coherence length and the expression $H_{c2} = \Phi_0/2\pi\xi^2$, where Φ_0 is the flux quantum, one can estimate H_{c2}. Here ξ is the Ginzburg–Landau coherence length:

$$\xi_{GL} = a\xi_0[1 - (T/T_c)^2]^{-1}, \tag{7.13}$$

where a in the weak-coupling approximation is equal to 0.74. Strong-coupling effects leads to an increase in a [29] and, for example, $a = 0.95$ for $\lambda \cong 2$. Using this value of a and $\xi_0 \cong 20\,\text{Å}$ we obtain the value of $H_{c2}(0) \cong 90$ T. It is important to note that this is the value that was used in order to evaluate $\gamma(0)$ (see above, Section 7.2.1) and obtain values of the major parameters (see Table 7.1), including the Fermi velocity v_F. The fact that we now obtain the same value of $H_{c2}(0)$ illustrates the self-consistency of our method.

A uniquely small value of the coherence length is a key feature of the cuprates. The small size of ξ_0 (see above) is directly related to the exotic values of the normal parameters.

7.3.2. The ratio $\varepsilon(0)/E_F$

The ratio $\varepsilon(0)/E_F$ is an important parameter in the physics of super-conductivity; it estimates what fraction of the carriers are directly involved in the pairing. This parameter is very small in conventional superconductors (10^{-4}). In the cuprates the situation is entirely different. A small value of the Fermi energy (along with a large value of the gap) leads to a large value of the ratio: $\varepsilon(0)/E_F \approx 10^{-1}$. This means that a significant fraction of the carriers are paired. Of course, this corresponds to a short average distance between paired carriers, which further implies a short coherence length.

The possibility of having a large value of $\varepsilon(0)/E_F$ and a short coherence length is directly related to the quasi-2D structure of the cuprates. Indeed, in conventional superconductors ($\varepsilon(0)/E_F \ll 1$) pairing can occur only near the Fermi surface (Cooper theorem). The states on the Fermi surface form a quasi-2D system in momentum space. This is an important factor because in the 2D case any attraction leads to the formation of bound states (see, e.g., [30]). The presence of the layered structure makes pairing possible even for states which are distinct from the Fermi surface; this corresponds to the picture in the cuprates.

7.3.3. Critical behavior

The large value of $\varepsilon(0)/E_F$ not only leads to a qualitatively novel picture of the pairing but also has a direct impact on a number of properties.

The ratio $\varepsilon(0)/E_F$ is directly related to the Ginzburg parameter [31] describing the scale of the critical region near T_c. The possibility of observing pair fluctuations (because of small value of F_F) has been predicted by Deutscher [32]; this effect is similar to that in liquid He-II. For example, instead of a sharp jump in heat capacity at T_c, one should observe structure similar to the λ-anomaly. Such an effect has indeed been observed experimentally [33]. In fact this led to an independent method for estimating the Fermi energy from the experimental width W of the critical region of the specific heat anomaly [7a]. The critical region can be written in the clean limit as

$$W < [(13/s)(k_B T_c/E_F)]^4, \tag{7.14}$$

where s is equal to $2\varepsilon(0)/3.52 k_B T_c$ or is proportional to the strength of the coupling. In fact, for YBaCuO this leads to an estimate for the renormalized E_F of 0.3 eV.

7.3.4. Positron annihilation

The small value of $\varepsilon(0)/E_F$ also leads to an interesting phenomenon with respect to positron annihilation in the cuprates and has been described in [9d]. It has been observed that the transition to the superconducting state leads to an increase in positron-annihilation lifetime [34] in the LaSrCuO compound and this shift, $\Delta\tau = \tau_s - \tau_n$, increases with decreasing temperature below T_c. This phenomenon is interesting because it has been observed only for the cuprates. A study of conventional superconductors did not reveal any shift in the lifetimes relative to the normal state; therefore, we are dealing with a phenomenon which is a unique feature of the new materials.

A positron entering the oxide annihilates mainly with localized electrons. However, its interaction with the localized electrons, which results in the annihilation, is screened by the delocalized carriers, and the screening appears to be noticeably affected by the pairing.

The annihilation lifetime is determined by the overlap of the positron and bound-electron wave functions [35]. Because of this, the expression for the lifetime can be written in the form

$$\tau^{-1} = \alpha \sum_n \int d\mathbf{r} \, \kappa_{ep}(\mathbf{r}, \mathbf{r}', \mathbf{r}_n)|_{\mathbf{r}'=\mathbf{r}}, \tag{7.15}$$

where $\alpha =$ constant and κ_{ep} is the two particle Green's function (the indices e and p correspond to the bound electron and the positron). The function κ_{ep} is determined by the Bethe–Salpeter equation, which contains the vertex Γ describing the positron–bound electron interaction. In RPA the vertex Γ has the form (see Eq. (4.13))

$$\Gamma(\kappa, \omega_n) = [V^{-1}(\kappa) - \Pi(\kappa, \omega_n)]^{-1},$$

where V is the direct Coulomb interaction and the operator Π describes the influence of the delocalized carriers (screening). The vertex Γ can be written in the form $\Gamma^{-1} = \Gamma_n^{-1} - \delta\Gamma$, where Γ_n corresponds to the normal state and $\delta\Gamma$ is the change in Γ caused by the pairing. The evaluation of $\delta\Gamma$ (for detailed analysis see [9d]) leads to the following shift in the lifetime:

$$\Delta\tau = \gamma\left(\frac{\varepsilon(0)}{E_F}\right)^2 \ln\left(\frac{E_F}{\varepsilon(0)}\right), \tag{7.16}$$

where γ is approximately unity. Therefore, the shift in lifetime depends directly on the parameter $\varepsilon(0)/E_F$. Note that the value of $\varepsilon(0)/E_F$ as a possible explanation of the shift was indicated also in [36].

For conventional materials the ratio $\varepsilon(0)/E_F$ is very small; one can obtain, for example, for aluminum, $(\varepsilon(0)/E_F)^2 \approx 10^{-9}$. Such a small shift is unobservable. Conversely, for LaSrCuO, the situation is entirely different: a large value of $\varepsilon(0)$ along with small value of the Fermi energy leads to an observable shift of several percent. The increase in the shift with decreasing temperature below T_c is due to the temperature dependence of the energy gap ($\Delta\tau \propto \varepsilon^2$). Therefore, positron annihilation measurements could be used in order to determine the temperature dependence of the energy gap.

7.3.5. Electromagnetic properties

The short coherence length also affects the electromagnetic properties. In fact, the unusual electromagnetic properties of the high T_c cuprates are due to several factors: (1) short coherence length ξ_0, so that $\xi_0 \ll \delta$, where δ is the penetration length; (2) anisotropy, presence of a layered structure; and (3) a two-gap spectrum (see below, Section 7.5). All these factors lead to behavior which differs in a drastic way from that in ordinary superconductors [37].

Consider a layered superconductor (Fig. 7.2) with its boundary perpendicular to the layers [11e] (*ab*-oriented film). In addition, we consider incident radiation normal to the surface of the film with the magnetic component perpendicular to the layers. Our goal is to evaluate the conductivity $\sigma(\omega, \kappa)$ and then the surface resistance describing the microwave

FIG. 7.2. High T_c superconductor in an a.c. field.

losses. The current is described by the expression:

$$j_y(\tau, \boldsymbol{\rho}) = \frac{ie}{m} [(\nabla_{\boldsymbol{\rho}} - \nabla_{\boldsymbol{\rho}'})G'(\tau, \boldsymbol{\rho}; \tau', \boldsymbol{\rho}')|_{\tau'=\tau+0}^{\boldsymbol{\rho}'=\boldsymbol{\rho}} - \frac{e^2 N}{mc} \mathbf{A}_y(\tau, \boldsymbol{\rho})\mathbf{n}] \quad (7.17)$$

where the summation is taken over all layers in a unit volume, $\boldsymbol{\rho}$ is a two-dimensional vector in the ab-plane, $\mathbf{n}$ is the unit vector along OY axis (OY $\perp$ A), and G' is the change in a thermodynamic Green's function due to an external field. The expression (7.17) can be reduced to the form

$$j_y(\omega, k) = -Q(\omega, k)A_y(\omega, k),$$

where

$$Q(\omega, k) = \frac{Ne^2}{mcL} - \frac{e^2}{16\pi^2 m^2 c} \int dp \, d\varphi \, p^3 \sin^2 \phi$$

$$\times \tanh \frac{E_+}{2T} \left\{ \frac{1}{E_+ - E_- - \omega - i\delta} + \frac{1}{E_+ - E_- + \omega + i\delta} \right\}.$$

Here

$$E_{+,-} = [\Delta^2 + \xi_\pm^2]^{1/2} \quad \text{and} \quad \xi_\pm = (1/2m)(\mathbf{p} \pm \mathbf{k}/2)^2 - \mu.$$

The conductivity is equal to

$$\sigma(\omega, k) = i(c/\omega)Q(\omega, k).$$

Evaluation [11e, 16] leads to the following expression (we assume $\hbar\omega < 2\Delta$):

$$Q_1 = \frac{n_s e}{mc}, \qquad Q_2 = f(T, d_c)\alpha \ln \alpha, \quad (7.18)$$

where

$$f(T, d_c) = \left(\frac{e^2 P_F^2}{\pi cm}\right)d_c^{-1} \frac{\varepsilon}{T} \cosh^{-2}(\varepsilon/T), \qquad \alpha = \frac{\omega}{k v_F},$$

and $n = n_s/d_c$ is the carrier concentration and n_s is the in-plane carrier density.

The calculation of the surface impedance (see e.g. [38])

$$Z = \frac{4\pi i\omega}{c^2} \frac{A(0)}{H(0)},$$

leads to the following result (we write out the expression for the surface resistance R_s describing the losses):

$$R_s = \frac{2}{\pi^2} \left(\frac{m\mu_0}{ne^2}\right)^{1/2} \frac{\Delta}{h} \left(\frac{\delta_0}{\xi_0}\right)^2 \gamma\left(\frac{l}{\delta_0}\right) \left[\left(\frac{\Delta}{k_B T}\right)^2 + \ln^2\left(\frac{\omega\delta_0}{v_F}\right)\right] \left(\frac{\hbar\omega}{\Delta}\right)^2 \left(\frac{\Delta}{k_B T}\right) e^{-\Delta/k_B T}$$

(7.19)

where l is the mean free path and

$$\gamma(x) = \frac{1}{x} [(1 + x^2)^{1/2} - 1]$$

is a universal function.

One can see that Z_1 contains a term $(\delta/\xi_0)^2 \gg 1$; this leads to large losses. Note also that there are finite losses even if $l \to \infty$, that is, in the absence of any collisions. Landau damping is a mechanism of such losses.

Note also that at $T \to 0$ $R_s \propto e^{-\Delta/T}$ and, therefore, $R_s \to 0$ if $T \to 0$. It is natural because $\hbar\omega < 2\Delta(0)$, and at $T = 0$ there is no absorption of the radiation. Nevertheless, experimentally one observes residual losses. Their appearance is partly due to inhomogeneities and second phases. However, in YBCO there is an additional channel for the losses due to the two-gap structure, and the appearance of an induced gapless state (see Section 7.5).

We have treated the case when the CuO layers are perpendicular to the boundary (*ab*-oriented film). A different case (*c*-oriented film) has been studied in [39]. In this case $v \perp k$ and the conductivity depends only on the frequency. The impedance displays a peculiar dependence on the mean free path. For example, there is a region where the conductivity decreases with an increase in the mean free path.

We have just described the electromagnetic response of a layered superconductor such as LaSrCuO. For YBaCuO, it is necessary to take into account the two-gap structure, and this will be described below.

7.4.   Induced superconducting state and two-gap structure

A small value of the coherence length along with the presence of different distinct structural units presents a unique opportunity to observe multigap structure in the cuprates.

In this section we focus our attention on YBaCuO, since this compound has both CuO planar structures and CuO chain structures, both of which are conducting and satisfy the necessary conditions to observe multigap physics.

In fact, this compound exhibits many unusual properties such as large residual microwave losses, zero-bias anomalies in the tunneling conductance, a plateau in the critical temperature versus oxygen content, a linear heat capacity at low temperatures, and a significant depression of T_c with praseodymium doping. These properties can all be understood in a unified fashion on the basis of the physics of induced superconductivity described in Section 6.4.

7.4.1. *Two-gap structure. Coherence lengths*

The YBaCuO compound contains a CuO quasi-one-dimensional chain structure in addition to the CuO planes. For the compound that is fully oxygenated ($Y_1Ba_2Cu_3O_{7-x}$, $x = 0$) the chain structure is particularly well developed. The chains provide doping for the CuO planes, but in addition and particularly important for our model, they form an independent conducting subsystem.

The conductivity of the chains has been demonstrated in several ways. The conductivity along the chain (b direction in the crystal) can be more than twice as large as in the a direction perpendicular to the chains [40]; this anisotropy indicates that the chains form a highly conductive subsystem. The same anisotropy has also been observed in the IR conductivity of untwinned single crystals [41]. Positron annihilation experiments have recently found the presence of planar parts of the Fermi surface that are a reflection of quasi-one-dimensional metallic structures [42].

Because there are two conducting subsystems and because their coherence lengths are uniquely short, there should be a two-gap-like excitation spectrum. Both the conducting chains and the two-gap structure were suggested in [6a, 11b] and have now been verified experimentally.

The existence of two gaps has been demonstrated by diverse experiments. The temperature dependence of the Knight shift [43] and NMR relaxation time [43, 44] for the plane and chain copper atoms was different and is the most direct evidence for two gaps, one associated with the planes and the other with the chains [7b]. Both the real and imaginary parts of the surface impedance are described by the sum of two contributions with different energy gaps [45]. Raman measurements on crystals of $YBa_2Cu_4O_8$ also show the presence of two gaps [46]. This compound has two planes and two chains and is thus very similar to $YBa_2Cu_3O_7$.

The fact that YBaCuO, with two conducting subsystems, has only one transition temperature means that there is pair transfer from one subsystem to the other; note that such charge transfer is an essential ingredient of our model described in Section 6.4. The doped CuO planes, which are the major structural and conducting unit for all the cuprate superconductors including LaSrCuO which contains only such planes, are by themselves intrinsically superconducting. The chains are intrinsically normal; this is supported by praseodymium substitution experiments [7b, 47].

Because of the presence of two conducting subsystems in YBaCuO, we want to apply our model [6e], see Section 6.4, to this compound; indeed, according to our approach the chains are intrinsically normal and their superconducting state is induced.

The double gap on the planes $2\varepsilon_\alpha(0)$ is equal to approximately $5T_c$. The value of the smaller gap $\varepsilon_\beta(0)$ is very sensitive to the oxygen content and for $x = 0$ is of the order of $1.25T_c$. This latter number has been obtained from careful measurements of the temperature dependence of the penetration depth at low temperature [48]. This value, which is also consistent with measurements of the temperature dependence of the surface resistance [45], is smaller than the BCS value of $1.76T_c$.

The application of the model described in Section 6.4 also leads to a number of peculiar properties of this compound. A decrease in the oxygen content greatly affects the value of the smaller energy gap, which has a very large impact on the spectroscopy of this compound. The reason for the dramatic effect on the small gap is the fact that the oxygen is removed preferentially from the chains. As a result, some of the chain copper atoms (see below) develop magnetic moments which act as a strong pair-breaker in the chain band. This causes the chain gap, ε_β, to be suppressed very rapidly, and the β subsystem becomes nearly gapless for oxygen contents around 6.9 (see Section 7.4.2). This shows up dramatically in spectroscopic experiments such as tunneling, IR reflection and absorption, photoemission and residual microwave losses.

According to Eq. (6.53), the gapless state for the β-subsystem arises if the mean free path l_β for the magnetic scattering is of the order of the coherence length ξ_β. For YBCO the index β corresponds to the chains, and α to the CuO planes.

Let us estimate the coherence length $\xi_\beta = \hbar v_F^\beta / \pi \varepsilon_\beta$ for the chains (for the planes $\xi_\alpha \approx 15$ Å). The Fermi velocity v_F^β is directly related to the density of states at the Fermi level, v_β, for a quasi-1D band, namely, $v_\beta = (\pi \hbar v_F a_0 L)^{-1}$, where a_0 is the interchain distance and L is the interlayer distance. According to band structure calculations [49], $v_\beta = r v_\alpha$, $r \approx 1.7$, where $v_\alpha = m_\alpha^* (\pi \hbar^2 L)^{-1}$ is the density of states for the planes. Thus, $\xi_\beta \approx \hbar^2 (\pi r a_0 m_\alpha^* \varepsilon_\beta)^{-1}$. Using the values $r \approx 1.7$, $a_0 \approx 3.5$ Å, $m_\alpha^* \approx 5m_e$ (see Table 7.1) and $\varepsilon_\beta \approx k_\beta T_c$ [28, 31], we obtain $\xi_\beta \approx 25$ Å. Experimental data [50] provide a different value of the density of states, namely, $r \simeq 1$. Then we obtain $\xi_\beta \approx 40$ Å.

Therefore, the coherence length on the chain is larger than that for the planes, but is still small relative to conventional materials.

7.4.2. Oxygen depletion and the gapless state

As we discussed above (Section 6.4.4), the presence of magnetic impurities in the β-group has a great impact on the energy gap ε_β, whereas T_c is not affected in a noticeable way. This picture is directly related to the $YBa_2Cu_3O_{7-x}$ compound in the interval $0.2 > x > 0$. Indeed, the removal of oxygen greatly affects the chain's states. First of all, some of the remaining oxygen atoms move to the interchain positions. As a result, we are dealing with the formation of a quasi-1D percolative structure. But, in addition, instead of a well-developed chain structure, we have a set of broken chains with copper atoms at the end. These copper atoms form local magnetic states, similar to surface states. These magnetic moments act as a strong pair-breaker in the chain band. As a result, the chain rapidly develops a gapless state.

It is easy to estimate the value of x which corresponds to the appearance of gapless superconductivity in the chains for random distribution of oxygen vacancies. Since $l_M^C \approx \xi^\beta$ (see above, Eq. (6.53)), and the value of ξ^β is between 25 Å and 40 Å (see Section 7.4.1), it is clear that the induced gap ε_β disappears for $x \approx 0.1$–0.15. Therefore, for $YBa_2Cu_3O_y$ with $y < 0.9$ we are dealing with gapless superconductivity on the chain.

The formation of the gapless state on the chain is very important for the spectroscopy of the material. The surface resistance, for example, is mainly determined by the smaller gap in the low-temperature region. A strong correlation between the oxygen content and the value of the smaller gap ε_β for random distribution of vacancies leads to the strong effect of the oxygen ordering on spectroscopy. As a result, one can observe significant residual losses for a sample with $x \gtrsim 0.1$. The quantum $\hbar\omega < 2\varepsilon_\beta$, where ε_β is the gap for the highly ordered sample ($x \approx 0$), can be absorbed for $x \gtrsim 0.1$, where the smaller gap ε_β disappears. On the other hand, the nearly stoichiometric oxygen content in the high-quality samples leads to a significant reduction in the residual losses. Such an effect has been observed recently [51].

The appearance of a zero-bias conductance (see, e.g., [52a]) is also a manifestation of the gapless state of the chains. Another consequence of this gapless state is a linear term in the heat capacity, which has been observed below T_c near $T = 0$ K [21b, 53].

Of course, the appearance of the gapless state of the chains does not mean that the material contains "superconducting" and "normal" components. The system still contains two superconducting subsystems and both order parameters are not equal to zero, but the β subsystem (chains) is in a peculiar gapless superconducting state. For example, this state displays the Meissner effect. Nevertheless, from the point of view of spectroscopy such terminology as "superconducting" and "normal"component can be used in a qualitative sense because the spectroscopic measurements do not find an energy gap.

Experimentally, one observes a plateau (or a small peak) in T_c versus x for $0.2 > x > 0$ [54, 55]. From the microscopic point of view, the critical temperature is determined by three parameters (λ_α, $\lambda_{\alpha\beta}$, $\gamma_{\alpha\beta}$, see above). The value of λ_α is determined by the doping level in the CuO planes, which is relatively constant in this region of doping [54]; on the other hand, the contributions from the other two parameters almost compensate each other [7b]. More specifically, the two additional contributions coming from the proximity channel (its contribution is proportional to $\gamma_{\alpha\beta}$) and the phonon-mediated interlayer channel ($\propto \gamma_{\alpha\beta}\lambda_{\alpha\beta}$) have opposite signs (see Section 6.4.2), and in $Y_1Ba_2Cu_3O_{7-x}$ they nearly compensate each other. The induced gap is drastically affected by the oxygen depletion becoming gapless even in this region of doping, but T_c changes very slowly ($\Delta T_c/T_c \simeq 5\%$).

The intrinsic proximity effect has also been studied [56]; however, the authors introduced only the pure proximity effect channel. In that case,

the conductivity of the chains would only depress T_c. The rigorous introduction of the phonon-mediated channel and the gapless state, as we have
done here, leads to an entirely different picture.

7.4.3. *Bi- and Tl-based cuprates*

In the BiSrCaCuO compound there are one, two, or three CuO layers
sandwiching two BiO layers. The compound can be made with differing
oxygen contents in the BiO layers without large changes in the transition
temperature. For those compositions with conducting BiO layers, the model
presented above is appropriate. Unfortunately, a detailed comparison with
the theory cannot be made since there is not enough information about the
induced gap, nor have there been enough studies of the properties of the
BiO layers, as is the case for YBaCuO. One can cite the result of a point
contact tunneling study [58] where there is a strong indication of the
presence of a small second gap.

With regard to the thallium-based cuprates, one can expect that the
TlO layer also forms a conducting subsystem. This follows from the
mixed-valence state of thallium observed experimentally [2, 59]. As a whole,
the problem of induced superconducting in the bismuth- and thallium-based
oxides deserves detailed study.

7.5.   Origin of high T_c

This section deals with the problem of the origin of high T_c in the cuprates.
The superconducting state in conventional superconductors is caused by the
electron–phonon interaction. At the same time it is known that pairing can
be mediated by other excitations. We discussed various mechanisms in
Chapters 2, 4, and 5.

We think that the high T_c in cuprates is due to strong coupling of the
carriers with low-energy phonons; at the same time the Coulomb repulsion
is weakened by the presence of an acoustic plasmon branch.

The key problem is the determination of the strength of the coupling.
Indeed, the critical temperature is determined by three parameters, the
strength of the coupling λ, the Coulomb pseudopotential μ, and the energy
scale (for the phonon mechanism, the energy scale corresponds to $\tilde{\Omega}$ of the
order of θ_D); so that $T_c = T_c(\lambda, \mu, \tilde{\Omega})$ (see Section 2.3.1).

If λ is small, then the high value of T_c cannot be explained on the basis
of the electron–phonon mechanism (λ could be small, for example, for the
excitonic mechanism; then the major benefit comes from a large energy scale)
because it would require too large an energy scale. Therefore, the phonon
mechanism requires strong coupling.

In principle, for conventional superconductors λ is determined by
tunneling spectroscopy (see Section 3.1), which is difficult to perform for the
cuprates because of the very short coherence length. Nevertheless, there has
recently been some progress (see below, Section 7.6).

We focus on alternative methods of determining the strength of coupling.

7.5.1. Determination of carrier–phonon coupling parameter

7.5.1.1. LaSrCuO

The determination of electron–phonon coupling is not a trivial task. In the absence of high-quality tunneling spectroscopy data, we present an alternative method based on the analysis of a fundamental bulk property, heat capacity.

Electron–phonon coupling, along with the temperature dependence of the phonon distribution, leads to the dependence of the Sommerfeld constant on temperature (see Section 2.5). This manifests itself in a deviation of the electronic specific heat from a simple linear law. At low temperatures ($T \to 0$) the Sommerfeld constant approaches the band value renormalized by the interaction (see Eq. (2.73)); this means that the carriers are "dressed" by phonons and other excitations.

We think that the major contribution to the coupling comes from the low-frequency optical modes ($\Omega \cong 10$ meV for LaSrCuO). Let us evaluate the coupling to this mode. We use the approach described in Section 2.5. One can see from Eq. (2.72a) and Fig. 2.5 that at $T = 0.25\tilde{\Omega}$ the carrier becomes undressed. The intensity of the thermal motion at T_c is sufficient to destroy the main part of the phonon "cloud," so that $\gamma(T_c) = \gamma_0$. Then we obtain

$$\lambda = \left(\frac{\gamma(0)}{\gamma(T_c)} \right) - 1. \tag{7.20}$$

Using Eq. (7.20), one can evaluate λ.

Thus, if we can use experimental data to extract the values of the low- and high-temperature Sommerfeld constants, then we can get an estimate for the minimum value of the electron–phonon coupling constant λ.

The value of $\gamma(0)$ can be obtained from the analysis of the dependence of γ on H (see above, Section 7.2). The value of $\gamma(T_c)$ can be obtained from the measurements of the jump, ΔC, in heat capacity at T_c and the expression $\Delta C/\gamma(T_c)T_c = \beta$, where β describes the jump in the strong-coupling theory (see Section 2.4, Eq. (2.70)).

As a result, we obtained λ to be in the range 2.2–2.5 [6c]. This means a strong carrier–phonon coupling (remember that lead, which is a conventional superconductor with strong coupling, has $\lambda \approx 1.4$).

7.5.1.2. YBCO

The situation in YBCO is more complicated than in LaSrCuO because of the presence of the chain band (see Section 7.4). This leads to the complications described in Chapter 6 with regard to induced superconductivity and multiple gap effects.

Although there are several coupling constants that need to be estimated —λ_α, the electron–phonon coupling constant for the CuO plane, $\lambda_{\alpha\beta}$, the coupling constant for the interband transitions, and $\Gamma_{\alpha\beta}$ which is related to the McMillan tunneling parameter—the most important one is λ_α because it plays the dominant role in determining the transition temperature whereas the terms involving the other parameters are compensatory and are important mainly for the spectroscopy. The technique that we used to estimate λ_α [13] is similar to the one just described for LaSrCuO since we base our determination on the ratio of the low temperature Sommerfeld constant to the high temperature value. However, the estimation of λ_α is more complicated than for LaSrCuO because of the existence of two conducting subsystems both of which contribute to the Sommerfeld constant. We have neglected the renormalization of the chain band because the coupling ($\lambda_{\beta\alpha}$ and λ_β) is small.

We determined the high temperature electronic specific heat by carefully measuring the total specifc heat from 300 to 700 K (see [60]) and then subtracting the lattice contribution including anharmonic contributions which we determined in part by integrating the phonon density of states. This is in contrast to the analysis of LaSrCuO where we estimated the unrenormalized Sommerfeld constant from the jump in the specific heat at T_c. The resulting electronic part was linear in T which gave us confidence in this result. The total Sommerfeld constant, $\gamma(0)$, was found to be approximately 25 mJ/mol-K^2 [13]. Using an experimental determination of the high temperature density of states of the plane carriers [50] we were able to estimate the separate plane and chain contributions to $\gamma(0)$. The plane contribution was 14 mJ/mol-K^2 and the chain contribution was about 11 mJ/mol-K^2. The low temperature value of the Sommerfeld constant was evaluated in much the same way as for LaSrCuO. We multiplied $d\gamma/dH$ measured at low temperatures with $H_{c2}(0)$ (cf. [61]) and found that $\gamma(LT)$ was approximately 60 mJ/mol-K^2. With use of [62, 63] we found $H_{c2}^{ave} = 171$ T. Subtracting the chain contribution from the low temperature value of γ we found that $\gamma_{pl}(LT)$ is approximately 49 mJ/mol-K^2. From the ratio of the high and low temperature values of the Sommerfeld constant we find that λ_α is at least 2.5. In this analysis we used very conservative estimates for all the quantities so the value of λ_α we obtained should be considered the minimum value consistent with the data. Note that this value is quite consistent with the value we determined the LaSrCuO and means strong electron phonon coupling for the plane carriers.

We have made estimates of the other coupling constants and parameters based on our interpretation of some crucial experiments. These results will be discussed in the next section where we use the estimates of the coupling constants to estimate T_c.

7.5.2. *Critical temperature*

The question of how high T_c can be due to the exchange of some particular excitation may be broken in two distinct parts: (1) the question of the existence of a particular excitation, and (2) whether the coupling of the carriers to this (or these) excitations is sufficient.

The answer to the first question for phonons is positive; we are talking about excitations that certainly exist. The existence of the layered plasmons has been justified theoretically and (see above, Section 7.2.2.2) and experimental verification is under way. As has been discussed in Chapter 4, the additional attraction caused by electronic excitations can be presented formally as a decrease in μ. Indeed, tunneling spectroscopy (see Chapter 3) will verify the presence of this additional attraction as a diminished or even negative value of μ. This picture is due to the fact that the static Coulomb repulsion as well as the dynamic part (plasmons) corresponds the energy scale of the electronic excitations, which is much higher than the phonon energies.

Let us discuss first the problem of T_c in the LaSrCuO compound. YBaCuO has a two-gap structure and will be discussed below.

The critical temperature for LaSrCuO can be estimated from Eq. (2.62). As was noted above, LaSrCuO is characterized by low frequency optical modes (see Section 7.2.2.1). As a result the characteristic (average) phonon frequency $\tilde{\Omega}$ is about 15 meV. Using a value of λ in the range 2.2–2.5 (see Section 7.5.1.1) and $\mu \approx 0$ we estimate T_c to be in the range of 35–40 K. Thus we can fully account for the superconductivity in the framework of a strong-coupled phonon attraction with a Coulomb repulsion weakened by a low-energy acoustic plasmon spectrum.

The situation inf YBCO is more complicated because of the two-gap structure. In this case there are several parameters to evaluate in order to estimate T_c but in fact the most important parameter is λ_α the intrinsic coupling parameter in the CuO planes. This parameter has been determined using measurement of the high and low temperature specific heats and is described in detail in Section 7.5.1.2. The value of 2.5 that we obtained is consistent with the value for LaSrCuO and indicated strong electron–phonon coupling.

The evaluation of T_c for YBCO is based on the equation

$$\Delta_\alpha(\omega_n)Z_\alpha(\omega_n) = \lambda_\alpha \pi T \sum_{n'} D_{nn'} \frac{\Delta_\alpha(\omega_{n'})}{|\omega_{n'}|} + \lambda_{\alpha\beta}\lambda_{\beta\alpha}(\pi T)^2 \sum_{n;n'} D_{nn''}D_{n''n'} \frac{\Delta_\alpha(\omega_{n'})}{|\omega_{n'}||\omega_{n''}|}$$

$$+ \lambda_{\alpha\beta}\Gamma_{\beta\alpha}\pi T \sum_{n'} D_{nn'} \frac{\Delta_\alpha(\omega_{n'})}{\omega_n^2} + \Gamma_{\alpha\beta}\lambda_{\beta\alpha}\pi T \sum_{n'} D_{nn'} \frac{\Delta_\alpha(\omega_{n'})}{|\omega_{n'}||\omega_{n'}|}$$

$$+ \Gamma_{\alpha\beta}\Gamma_{\beta\alpha} \frac{\Delta_\alpha(\omega_n)}{\omega_n^2}. \tag{7.21}$$

and the corresponding equation for the renormalization function Z. These equations follow directly from Eqs (6.23–6.26). In fact, Eq. (7.21) is more general than Eqs (6.29) and (6.35) which are applicable only to limiting cases.

The parameters which contribute to the critical temperature in addition to λ_α are $\lambda_{\alpha\beta}$, $\gamma_{\alpha\beta}$, $\tilde{\Omega}$, and μ^*. Each of these parameters has a definite physical meaning and can be estimated from various experimental measurements. Measurements of the penetration depth [48] and microwave surface resistance [45] have allowed us to estimate the induced energy gap ε_β and from its value estimate $\gamma_{\beta\alpha}$ to be of the order of 0.3. Using band structure calculation estimates for the ratio of densities of states in the plane and chain bands we have estimated that $\gamma_{\alpha\beta}$ is approximately 0.5. From measurements of the resistivity in the b (chain) direction in untwinned crystals we have estimated that the interband coupling constant $\lambda_{\beta\alpha}$ is about 0.5 and that $\lambda_{\alpha\beta}$ is about 0.8. Based on data [50] we obtain slightly different values: $\gamma_{\alpha\beta} \simeq 0.3$, $\lambda_{\alpha\beta} \simeq 0.5$.

The average phonon frequency $\tilde{\Omega}$ has been estimated from the phonon spectrum measured by neutron scattering and is approximately 330 K.

Finally, the presence of a low frequency acoustic plasmon mode common to all the cuprates [9b,c] (see above) leads to a dynamic effect and also to an additional attraction. This effect is due to electron–electron correlations and corresponds to the same energy scale as the static attraction. As a result, the presence of the plasmons can be treated as a factor leading to weakening of the Coulomb repulsion and effectively diminished μ^*. The result is a μ^* that is in the range of 0.0 to -0.1. Using this set of parameters and the matrix method [64] we have solved Eq. (7.21). The resulting value for T_c is in the range of 80–90 K. One important note is that the interband coupling constant and the internal proximity affect the transition temperature in opposite directions and nearly compensate each other for the parameters that we have determined. Thus the value of the transition temperature using a single band Eliashberg equation would be about 80 K for a μ^* of the order of zero.

7.5.3. Discussion

We have accounted for the experimental T_c in these compounds on the basis of a strong electron–phonon coupling parameter, λ, in the range of 2–3. These values are not extraordinary and in fact there are lead–bismuth alloys that have comparable λ values [see e.g. 57a]. Thus, the phonon interaction can give rise to very high transition temperatures, contrary to some conventional (but incorrect) wisdom. There is no contradiction with the present theory (see Chapter 2).

In fact, the importance of phonons in cuprates is usually minimized because of the small value of the oxygen isotope effect. However, the isotope effect, particularly for complicated compounds, is not at all a simple and straightforward phenomenon (see Section 2.7). It is well known that many conventional superconductors have a very small isotope effect even though the mechanism of superconductivity is definitely phononic. The presence of a polyatomic lattice [65], the effects of the isotope substitution on the elastic constants, a plasmon contribution, etc., lead to a situation where it is inappropriate to correlate the value of the isotope coefficient with the contribution of the phonons to T_c. Furthermore, the isotope coefficient is not small for the nonoptimally doped material (see below, Section 7.6).

Another argument against phonons is connected with the evaluation of λ from normal transport data [52b]. However, this analysis dealt with the so-called transport coupling constant λ_{tr} which is related to but nevertheless different from the λ entering T_c. According to [57b], the presence of a layered structure leads to a large difference between λ and λ_{tr}. In addition, it is necessary take into account the anisotropy of the system and the two-band structure, particularly the conductivity of the chains in a fully oxygenated sample of YBaCuO, and finally, but no less importantly, a small value of E_F (see Section 7.2.2). This latter fact requires that the temperature dependence of the chemical potential $\mu(T)$ be considered in the high-temperature region. All these factors affect the value of λ.

Note that the phonon-mediated superconductivity, contrary to magnetic mechanisms (d-wave), assumes s-wave pairing. Therefore, the determination of the symmetry of the pair function is playing a crucial role. In connection with this, one should note that, although, there are some data which are in an agreement with both, s-wave as well as d-wave symmetries, there is a strong and growing experimental support in favor of s-wave pairing. We will discuss this question in Section 7.6.2.

7.6. Key experiments

In this section we will discuss many crucial experiments. We will also include a discussion of some experiments whose interpretation is still not clear but which are important for a complete understanding of these very interesting compounds. This section is organized to follow the progression of the earlier part of this chapter.

7.6.1. Normal properties

7.6.1.1. Fermiology

Since the superconducting cuprates are doped insulators, it was not clear a priori that they could be described by the Fermi liquid picture, which means that they had a Fermi surface and that the language of Fermiology was appropriate to describe the normal state. That this was indeed the case was first demonstrated in 1987 [6a,b] on the basis of specific heat measurements [21] which found a linear low-temperature specific heat that in addition varied linearly with magnetic field. Magnetic field creates a normal phase (vortex region). The linear dependence $C_{el} \propto T$ is a clear indication of Fermi liquid behavior [19]. More recently several other experiments have confirmed the existence of a Fermi surface and in fact have determined some of its specific features. Photoemission data on both YBaCuO and BiSrCaCuO show clear Fermi edges [66]; furthermore, angle-resolved data have looked in detail where the various bands cross the Fermi surface and found good agreement with predictions of the LDA band structure calculations [49]. Angle-resolved positron annihilation experiments have now found convincing evidence for the sheetlike pieces of the Fermi surface that arise from the chain band in YBaCuO [42a], demonstrating that the chains are indeed a well-defined conducting subsystem. The authors [42a] performed their measurements on twin-free crystals of $YBa_2Cu_3O_{7-\delta}$. The electron–positron momentum density displays the presence of the "ridge" Fermi surface sheet. Later the signal corresponding to the Fermi surface was isolated for the twinned crystal [42b]. It is important that the signal is absent in the insulating phase; therefore, this is a property of the doped state. Finally, de Haas–van Alphen measurements at several laboratories [67] have found oscillations characteristic of a small hybrid piece of the Fermi surface, again predicted by the band structure calculations. Thus, there is no longer any doubt that these materials indeed can be described by the parameters of Fermiology and that they have a well-defined Fermi surface, although the problem of complete reconstruction of the Fermi surface still remains.

7.6.1.2. Phonon spectrum

The phonon spectrum and its dispersion are very important features of the collective excitation in the normal state. In fact, comparison of the $F(\Omega)$ determined by neutron scattering with $a^2(\Omega)F(\Omega)$ determined from tunneling spectroscopy (see Chapter 2) was instrumental in proving that most conventional superconductors are phonon-mediated. For the cuprates, neutron scattering measurements [24] have found that they all have soft optical phonons in the 10–40 meV energy range that have the appropriate symmetry to couple strongly to the carriers. Since tunneling spectroscopy is not yet mature in the cuprates (see below), neutron scattering measurements have taken on a more crucial role in the understanding of the nature of the superconducting state in the cuprates.

7.6.1.3. Transport measurements

Transport measurements have provided many insights into the properties of the cuprates. The nearly linear resistance above the transition [68] extending to 1000 K for LaSrCuO and to 700 K for YBaCuO has caused much speculation as to the nature of the normal liquid. In fact, the linear resistance is well explained by electron–phonon scattering provided one takes into account the details of the phonon density of states, anisotropy and the small Fermi energy [69]. Furthermore, the large in-plane anisotropy of the resistance in untwinned single crystals of YBaCuO is evidence for the existence of separate conducting plane and chain bands. Moreover, recent results have shown that the conductivity along the a direction (perpendicular to the chains) is is relatively insensitive to the oxygen content in the range 6.8–6.95 whereas the conductivity in the *b* direction (along the chains) is drastically modified in this same range, indicating the dramatic influence of oxygen content on the chain band [70].

It was recognized very early in the study of the cuprates that the thermal conductivity in the normal state was dominated by phonon transport rather than electronic transport as is the case with conventional metals [71]. This is strong evidence for a small value of the Fermi energy as we estimated above (see Section 7.2.1). In fact, the strong upturn in the thermal conductivity below the transitionto the superconducting state is indicative of an increase in the phonon mean free path due to the reduced phonon–electron scattering as the pairs condense into the superconducting state [72].

Another feature of the normal state that is quite interesting is the magnetic susceptibility measured either by a SQUID or by NMR. For the optimally doped compounds, the susceptibility is temperature-independent and gives a reasonable estimate for the unrenormalized density of states [73a]. However, in YBaCuO as the oxygen content is reduced below the ideal value of 7.0, the susceptibility begins to show a negative slope starting from progressively higher temperatures as the oxygen content is slowly lowered [73b]. This is still an unexplained result and may be an important feature of the normal state.

7.6.2. *Superconducting properties*

Again many of the features of the superconducting state that we have discussed in the earlier sections of this chapter have been observed experimentally. This section will describe several interesting results.

7.6.2.1. *Energy gap and two-gap structure*

The observation of a clear energy gap or energy gap structure has been quite controversial, but there are a number of recent experiments in which the results are converging and a consistent picture is emerging (see the review [74]). Some of the earliest results on the observation of separate plane and chain gaps were the NMR measurements of both the AT&T [44] and Illinois [43] groups. They observed a large gap associated with the planes $(2\varepsilon(0)/k_B T_c \approx 5)$ and a smaller gap associated with the chains $(2\varepsilon(0)/k_B T_c \approx 3)$. These results were among the earliest and best confirmations of the two-gap model that we discussed above (see Section 6.1). However, the magnetic field dependence of the anisotropy of the relaxation times [75] is still an open question and may bear on the symmetry of the pairing (d- or s-wave) or on the condition of the chain band which could be gapless (see Section 7.4.1).

Tunneling, as discussed in Chapter 2, has been the most reliable method for determining the energy gap as well as the nature of the pairing interaction in conventional superconductors. For the cuprates, the situation is complicated by the very small coherence length, and in the case of YBaCuO, BiSrCaCuO, and TlBaCaCuO by the presence of multiple-gap structures and induced superconductivity (see Section 7.4). The current status is that multiple-gap structures have been observed clearly in YBaCuO [52] and BiSrCaCuO [76], with several independent measurements of the large gap for YBaCuO consistent with the NMR results, i.e., $2\varepsilon(0)/k_B T_c \approx 5$. Recent point contact proximity tunneling results also on YBaCuO [77] have found two gaps that are almost identical to the NMR values whereas most of the other tunneling experiments on YBaCuO have found smaller values for the second gap [78]. The value of the large gap for BiSrCaCuO is somewhat larger than the corresponding YBaCuO gap and appears to be in the range $2\varepsilon(0)/k_B T_c \approx 6-7$. The smaller gap has not been consistently observed, but when it has it has been very small, i.e., $2\varepsilon(0)/k_B T_c \approx 1$ [78]. Overall, the tunneling experiments seem to show multiple-gap structures in the higher T_c compounds.

Infrared measurements in the superconducting state on untwinned single crystals have shown a clear anisotropy of the reflection [79a] and absorption [79b] between the a (perpendicular to chain) and b (along chain) crystalline directions. Along the a direction there appears to be a very large gap ($2\varepsilon(0)/k_B T_c \approx 8$) with some very small residual absorption, whereas along the b direction there does not appear to be a gap at all. Our interpretation of these results is that the single crystals are not well oxygenated and are in the gapless chain regime. In this case there would be states in the gap in both orientations due to the finite charge transfer between them.

Microwave surface resistance measurements on films and crystals have in general found large residual losses at low temperatures consistent with unpaired carriers at low temperatures [80]. Very recently there have been measurements on high-quality, well-oxygenated films that showed an exponential decrease of the surface resistance at low temperatures consistent with the presence of a gap [51]. These measurements are strong evidence that *all* the carriers are paired at low temperatures. Penetration depth measurements, also at microwave frequencies, have shown an indication of a two-gap structure; a large gap manifests itself near T_c, whereas a small gap dominates the exponential approach of the penetration depth to its low-temperature value [48].

Raman measurements have also shown a clear indication of multiple-gap structures, and in particular have found two gaps in $YBa_2Cu_4O_8$ [46]. These gaps are very similar to the gaps found in the $YBa_2Cu_3O_7$ compound by the NMR measurements described above.

7.6.2.2. Symmetry of pairing

An important issue, and one that is closely related to the mechanism of superconductivity in the cuprates, is the symmetry of the superconducting ground-state wave function. All phonon-mediated superconductors have wave functions with s-wave symmetry and therefore have a gap that is nonzero everywhere on the Fermi surface. Magnetic pairing interactions, on the other hand, have higher-symmetry wave functions with zeros in the gap. The manifestation of zeros in the gap would be very similar to gapless superconductivity, i.e., the presence of unpaired carriers at very low temperatures which would occur because of oxygen vacancies in the chain band (see above). This would give rise to such anomalies as high residual microwave losses and zero-bias conductance in tunneling measurements. Thus the recently measured exponentially decreasing surface impedance for a fully oxygenated sample [51] and exponential approach of the penetration depth to its zero-temperature value [46] discussed above seem to strongly favor the s-wave ground state. Note that a power law dependence can arise from interactions of quasiparticles with low frequency phonons [46b].

Measurements of persistent currents and transition temperatures [81] in composite superconductors consisting of a cuprate and a conventional superconductor (lead) have been interpreted as a clear indication of an s-wave ground state.

Measurements of the temperature dependence of the anisotropy of the NMR relaxation times in untwinned single crystals [75] has been interpreted as evidence for d-wave pairing [82]. An alternative explanation for this dependence could be a gapless state of the chain band (see Section 7.4). The gapless state is characterized by nonexponential, but linear, temperature dependence of the electronic heat capacity. The picture is equivalent to the presence of nodes in the order parameter.

Although there is not yet a consensus on the symmetry of the superconducting wave function, it appears that the majority of evidence is in favor of a conventional s-wave ground state.

7.6.2.3. Estimates of λ

The clearest indication of the applicability of the electron–phonon mechanism would be a tunneling determination of λ. There have, of course, been many attempts at tunneling spectroscopy and the current situation is hopeful, but not yet completely convincing because of the difficulties associated with making a good tunnel junction (see [74] for a good review of the state of the art). Point contact spectroscopy on lower T_c members of the cuprates family has to date provided some of the most convincing results. For $(NdCe)_2CuO_4$ a cuprate with a transition temperature just over 20 K, a complete Rowell–McMillan inversion has been performed, and $a^2F(\Omega)$ and λ have been determined that gave nearly the measured T_c [83]. The $a^2F(\Omega)$ was very similar to the $F(\Omega)$ determined by neutron scattering and thus provides convincing evidence for the phonon mechanism for this compound.

For the higher T_c members of the cuprate family, heat capacity measurements as discussed above (Sections 7.5.1.1 and 7.5.1.2) have provided the best estimates of λ for LaSrCuO and YBaCuO.

7.6.3. *Properties as a function of doping*

Some of the most relevant experiments that are consistent with our viewpoint are concerned with the variation of the normal and superconducting properties as a function of the doping as provided by the replacement of La by either Sr, Ba, or Ca in La_2CuO_4 or by the removal of oxygen from $YBa_2Cu_3O_{7-x}$.

7.6.3.1. *Resistivity, thermopower, and T_c plateau in YBaCuO*

As was mentioned in Section 7.6.1.3, the anisotropy of the resistivity in untwinned crystals of YBaCuO has provided strong evidence for the two-band structure of this compound. Recent measurements of the resistivity and thermopower as a function of oxygen content in untwinned single crystals has clearly demonstrated that the transport properties in the *b* (chain) direction are severely affected by small modifications of the oxygen content, whereas the transport perpendicular to the chains is hardly affected [70]. Furthermore, in earlier studies [12] of the thermopower as a function of oxygen content in twinned single crystals it was found that the thermopower underwent a drastic change from strongly negative for oxygen contents near 7.0 to strongly positive for oxygen contents about 6.8. In this region of oxygen stoichiometry, the transition temperature remained above 90 K (this region of relatively constant or slightly peaked T_c was first discovered by Cava [54a]). This behavior can be understood if we ascribe the changes in the transport as due to changes in the chain band. As we remove oxygen from YBaCuO, it comes exclusively from the CuO chains. The chains lose their continuity as well as a large fraction of their density of states as the oxygen content goes from 7.0 to 6.8. Thus the thermopower is dominated by the carriers of the chains at high oxygen contents but is dominated by the hole carriers of the planes for the oxygen 6.8 samples. Since the transition temperature is relatively constant, this strongly supports the view that the chains by themselves are not crucial in providing the superconductivity, but are essential in understanding the transport and thermodynamic properties of the normal state and because of their induced superconductivity are crucial in understanding the spectroscopy of the superconducting state (see Section 7.6.2.1).

In fact, the very weak dependence of the transition temperature on *x* in the range 6.8 to 7 is a consequence of the fact that in this region only the doping of the chain band is affected, the doping of the CuO planes remaining nearly constant [54, 55]. However, since the chains are in an induced superconducting state that is very sensitive to the properties of the chain band, many of the important characteristics of the overall superconductivity other than T_c are drastically modified. In particular, any spectroscopic measurements (e.g. tunneling, microwave, infrared, photoemission, etc.) are drastically changed, especially if the chain band becomes gapless (see Section 7.6.2.1).

7.6.3.2. T_c versus carrier concentration

High T_c oxides are all doped materials and that is why the dependence of various parameters, especially T_c, on the carrier concentration deserves special attention. According to an experimental study [84a], T_c increases with increase in the strontium content in La_2CuO_4, and one can in fact observe a maximum in T_c. A similar effect has been also observed for bismuth-based cuprates [84b, 85].

One can show [9e] that the detailed analysis of the electron–phonon coupling leads to a nonmonotonic dependence $T_c(n)$. The appearance of the maximum in T_c is due to crossover between the Fermi momentum p_F (for layered conductors $p_F \propto n^{1/2}$) and the phonon momentum q_c of the optical mode ($\Omega = \Omega_1$ up to $q = q_c$). The coupling constant λ can be written in the form

$$\lambda = p_F^{-1} h(p_F); \qquad h(p_F) = \frac{2\pi m}{d_c} \int_0^{k_1} \frac{dq_{\parallel} \, |\xi}{[1 - (q_{\parallel}/2p_F)^2]^{1/2}}.$$

Here $\mathbf{q}_{\parallel} = \mathbf{p} - \mathbf{p}^1$ is a transfer momentum, ξ is a matrix element, and $\kappa_1 = \min\{2p_F, q_c\}$. If the carrier concentration is small so that $2p_F < q_c$, then $\kappa_1 = 2p_F$ and $\lambda \propto n$. Then λ, and therefore T_c, increases with increasing n. A further increase leads to a situation where $2p_F > q_c$, then $\kappa_1 = q_c$. This crossover leads to a change in the dependence $\lambda(n)$. For example, if $2p_F \gg q_c$, then $h \propto n^{-1/2}$. Therefore, T_c has a maximum at $n \cong q_c^2(8\pi d_c)^{-1}$ ($\cong 40$ K for LaSrCuO).

The nonmonotonic dependence $T_c(n)$ has been observed before in superconducting semiconductors (see, e.g., [86]). This effect has been explained in [87] by the interplay between the electronic and phonon momentum. It is important to note that the appearance of a maximum in T_c as a function of n can be explained in the framework of the electron–phonon interaction.

7.6.3.3. Carrier concentration and properties of the cuprates

We discussed above a strong dependence of T_c on carrier concentration. Carrier concentration appears to be a very important parameter which affects not only T_c, but many other properties of the materials, among them the isotope effect (see Section 7.6.3.4) and photoinduced superconductivity [88]. The latest effect displays a shift in T_c caused by photoexcitation. The oxygen stoichiometry was well defined and controlled [88c]. Photoexcitation led not only to a noticeable shift in T_c but also to a decrease in the normal resistance of insulating films.

A strong field effect has been observed in [89]. An electric field applied to the surface of the cuprate affects the carrier concentration and leads to a noticeable shift in T_c. A particularly large shift has been observed for a sample with $n < n_{c\,max}$. This effect is interesting from the point of view of future applications in electronics.

The temperature dependence of the normal resistance which dramatically depends on the level of the doping has been observed in LaSrCuO [90]. The linear dependence corresponds to optimally doped samples when $T_c \approx T_{c\,max}$. Deviations from stoichiometry lead to a different dependence.

In all these data as well as the µSR measurements [91] which also correlate T_c and n, the carrier concentration n is manifested as an important parameter. A phenomenological description of this dependence is presented in [92], where it was shown that the scaling ansatz $\bar{T}_c = \bar{n}(1 - \bar{n})$, where $\bar{T}_c = T_c/T_{c\,max}$; $\bar{n} = n/n_{max}$ can describe, in a universal manner, the cuprates and also the Chevrel phase systems.

7.6.3.4. Isotope effects

Early oxygen isotope experiments found very small shifts in LaSrCuO [93a] and negligible shifts for YBaCuO [93b]. These were used to argue strongly against the conventional phonon picture. In fact, there are many phononic superconductors that have small or even negative isotope shifts [94]. The appearance, size, and sign of the isotope effect is a very complicated problem that has been discussed in detail in Chapter 2, Section 2.7. More recent results have found a very strong dependence of the shift on the doping in LaSrCuO [95] and YBaCuO [96]. As the doping shifted away from the optimal for maximum T_c (on the underdoped side), the value of the isotope coefficient rose dramatically. Models based on the presence of a van Hove singularity at the Fermi surface [97] have been proposed to explain these results. This model should be supported by the observation of these singularities at E_F.

One can give a very different explanation [98]. We are dealing with a peculiar situation. A maximum value of T_c corresponds to a minimum value of the isotope coefficient α. It is important to realize that we are specifically dealing with the isotopic substitution $O^{16} \rightarrow O^{18}$ of the apical oxygen. The effect of this substitution strictly speaking is two-fold. First, it may affect the lattice dynamics, and this corresponds to the usual isotope effect. However, since we think that the major contribution to the pairing comes from the soft optical mode (see Section 7.5), and oxygen motion does not contribute significantly to this mode [99], oxygen motion by itself corresponds to higher frequencies. Therefore, the effect of this substitution on an important phonon frequency is small. Nevertheless, in a system like YBCO there is another effect of the isotopic substitution. Namely, the substitution affects the relative position of the apical oxygen. As a result, the doping, and, hence, the carrier concentration are affected. Such a scenario forces us to focus again on the important parameter, namely, on the carrier concentration (see previous section). Then the observed behavior of the isotope coefficient can be understood. Indeed, the dependence $T_c(n)$ has a maximum at some value $n = n_{max}$. The isotope coefficient which is proportional to the derivative $\alpha \propto \partial T_c/\partial M \propto (\partial T_c/\partial n)(\partial n/\partial M)$ should have an almost zero value, if the function (in our case $T_c(n)$) has a maximum. The microscopic picture is supported by a phenomenological description [92]; according to [92] the scaling ansatz $T_c(n)$ allows us to describe the dependence of α on n.

7.7. Organics vs. cuprates

7.7.1. *Why is T_c still so low in the organics?*

Organic superconductors represent a relatively young family of superconducting materials. The phenomenon of organic superconductivity was predicted in 1964, and the first such material was synthesized in 1980 (see Chapter 1). Since the discovery, there has been remarkable progress in the field and at the present time there are many organic superconductors which can be classified in several groups [100]. The recent discovery of the C_{60}-based superconductors (by definition they belong to the organics) with T_c close to 30 K made the derivative dT_c/dt (t is time) very impressive. Some of the organic materials are remarkably similar to the cuprates. Below, we are going to focus on this similarity, and, in addition, discuss the mechanism of superconductivity in the fullerenes.

Remarkable progress in the field of organic superconductivity has demonstrated the great potential of these materials for future development. Therefore, it is of definite interest to understand the origin of the present limitations of T_c. We think that a comparison of the organic superconductors and the high T_c oxides [101] is a fruitful approach to this problem. The organic superconductors which belong to the $(ET)_2X$ family (e.g., (BEDT-TTF)$_2$ Cu(SCN)$_2$, which has one of the highest values of T_c amongst the organics) and high T_c cuprates have a lot of similarities and some differences.

Both classes of materials are characterized by a large anisotropy. $(ET)_2X$, where ET is the abbreviation for BEDT-TTF, have a quasi-layered structure; the same property turns out to be a common feature of cuprates and it is one decisive factor determining their high T_c. In addition, both classes have a relatively small and quite similar carrier concentration. The evaluation is based on Fermiology. For the LaSrCuO compounds, the Fermi surface is cylindrically shaped. The Fermi surfaces of $(ET)_2Cu(SCN)_2$ ($T_c \cong 10.4$ K) [102] and $(ET)_2KHg(SCN)_4$ ($T_c < 0.5$ K) [103] are also neary cylindrical. Their parameters have been obtained by using conventional methods (Shubnikov–de Haas oscillations, and magnetoresistance); the ability to use these methods is due to a relatively low value of T_c and upper critical field (one can study the normal state in the low temperature region). One can estimate the Fermi velocity, v_F, the Fermi energy, E_F, and the Fermi wave vector, k_F, for $(ET)_2Cu(SCN)_2$ from the upper critical field H_{c2} by the following equations:

$$v_F = 4.4\Delta(0)/h[\phi_0/2\pi H_{c2}]^{1/2}$$

$$E_F = m^* v_F^2/2 \quad \text{and} \quad k_F = m^* v_F/h$$

where $\Delta(0)$ is the zero temperature energy gap, ϕ_0 is the flux quantum and m^* is the measured effective mass. We used a value for $\Delta(0)$ of $4.0 k_B T_c$ [104]. As a result we obtain for $(ET)_2Cu(SCN)_2$ the values $m^* = 1.4 m_e$, $E_F = 0.05$ eV, $v_F = 10^7$ cm/sec; for $(ET)_2KHg(SCN)_4$, $m^* = 3.5 m_e$, $E_F = 0.07$ eV, and $v_F = 10^7$ cm/sec.

According to [102], the $(ET)_2KHg(SCN)_4$ material has a large in-plane anisotropy of the normal conductivity (1:2). This anisotropy is probably present for the $(ET)_2Cu(SCN)_2$ compound as well. The CuO sheet in the cuprates does not display a strong anisotropy although YBCO has an anisotropy in the a-b plane due to the contribution of the chains. The presence of the anisotropy in the organics means that the cross-sections of the cylindrical Fermi surfaces of $(ET)_2X$ and LaSrCuO cut by the plane $P_z = $ const. have a different geometry. Namely, the cross-section for $(ET)_2X$ is a stretched ellipse, whereas for LaSrCuO the deviation from circular shape is not large. The presence of the stretched ellipse means the presence of nesting states and this is an important factor which affects the value of T_c (see below).

Either using values of the Fermi velocities or the upper critical field H_{c2}, one can estimate the values of the coherence lengths in these materials. ξ_0 for the cuprates is short (~ 15–25 Å). It turns out that the value of ξ_0 for $(ET)_2Cu(SCN)_2$ is also short ($\cong 70$ Å) relative to conventional super-conductors. At the same time, the mean free path in the organic super-conductor is large ($\cong 340$ Å, see [102]). Therefore, the criterion $\xi_0 < l$ is satisfied; it means (see Section 6.1), that in the presence of the overlapping energy bands, one can observe two gaps. A very similar situation occurs in the YBCO compound (see Section 7.4).

Note that the analysis of the band structure in $(ET)_2X$ [105] shows that two bands cross the Fermi level. Moreover, tunneling data [106] indicate the presence of two gaps in a similar organic compound. As a result, we think that the presence of the two-gap structure along with the anisotropy of each gap is perfectly realistic.

Secondly, a number of experimental data [107], reveal temperature dependences of several quantities (e.g., spin susceptibility) that are different from the usual BCS exponential dependence. This has been interpreted as a manifestation of an unconventional nonphononic mechanism (probably magnetic) of superconductivity. However, the presence of multigap structure allows us the opportunity to give a different interpretation. In this case, the temperature dependence is a sum of exponentials, and such a can match the observed power law.

A large value of the effective mass in the LaSrCuO compound is connected with strong coupling between the carriers and lattice (see Section 7.5.1.1). The value of the effective mass, m^*, in $(ET)_2KHg(SCN)_4$ is 1.4 and 3.5 for $(ET)_2Cu(SCN)_2$, whereas for LaSrCuO, $m^* \cong 5$. This difference is related to the strengths of the electron–phonon coupling. By comparing the measured effective masses of these two organic compounds the first has a

$T_c < 0.5$ K and the second a T_c of 10.4 K. We can make the simple assumption that the enhanced effective mass of the superconducting compound is due to a much larger electron–phonon coupling constant. By setting the m^* of $(ET)_2KHg(SCN)_4$ equal to the m^b of the high T_c compound and using the equation $m^* = m^b(1 + \lambda)$ we estimate that λ for $(ET)_2Cu(SCN)_2$ is approximately 1.5. This is indicative that the organics are strong coupled but not very strong coupled as we found for LaSrCuO. Indeed, a value of 1.5 for λ can adequately account for the measured transition temperature of 10.4 K.

We noted above that the Fermi surface of the organic superconductor contains a large number of nesting states. This is very favorable for the occurrence of a charge density wave (CDW) transition in this material. A CDW has the effect of opening up an energy gap over the nested parts of the Fermi surface, preventing this phase space from forming Cooper pairs. Qualitatively, this means that the effective dimensionality of the high T_c organic superconductor is intermediate between quasi-2D and quasi-1D, in contast to the much more quasi-2D character of the Cu–O planes in the cuprates. Again, the presence of the nesting states in this case is a negative factor [108] because of the transition at $T > T_c$ to the CDW state. We think that this factor (intermediate dimensionality between quasi-1D and quasi-2D) is important and its elimination might lead to further increases in T_c of organic low-dimensional materials.

We have shown that the cuprates and organics are very similar in their normal state properties differing mainly in their smaller values for the Fermi energy and **k** vector. This reduced phase space for pairing is due to their intermediate dimensionality (between quasi-1D and quasi-2D) and nested Fermi surface.

7.7.2. Superconducting fullerenes

A new family of superconductors (C_{60}-based materials) was discovered in 1991 [109–112]. These materials are fascinating molecular crystals whose lattices contain soccer-ball-like C_{60} clusters. The values of T_c are also impressive. For Rb_3C_{60}, $T_c = 28$ K; this greatly exceeds the critical temperatures of all conventional superconductors. Isolated clusters were identified in [113], and this was followed by the discovery of a method for the preparation of macroscopic amounts of the material [114]. Samples become metallic when alkali metal atoms are intercalated in the interstitial sites [115].

There has been considerable progress in the study of various properties of superconducting fullerenes (see, e.g., [116]). Here we focus on the question of the electron pairing mechanism in these materials.

Experiments involving the substitution $C^{12} \to C^{13}$ revealed a large isotope effect; for example, the isotope coefficient, α, equals 0.37 in Rb_3C_{60} [117]. Thus it has become clear that we are dealing with a phonon-mediated superconducting state. Nevertheless, there remains the very important question of identifying the vibrational modes responsible for the pairing. A related question is the determination of the coupling strength λ and the Coulomb pseudopotential μ^*.

The vibrational spectrum contains a variety of modes (see, e.g., [116, 118]). Among them are intermolecular modes ($\Omega_{inter} \lesssim$ 30–40 cm^{-1}), and intra-molecular vibrations. The latter include both low-frequency ($\Omega_L \lesssim$ 250–500 cm^{-1}) and high-frequency parts ($\Omega_H \cong 1.5 \times 10^3$ cm^{-1}).

It is also important to stress that the superconducting fullerenes are characterized by a small value of the Fermi energy (see, e.g., [116, 119]). For example, $E_F \cong 0.2$ eV in Rb$_3$C$_{60}$, $E_F \cong 0.3$ eV in K$_3$C$_{60}$. This feature is similar to that in the organics (see Section 7.7.1) and in the cuprates (Section 7.2.1). As a result, a description based on the weak-coupling BCS model cannot be expected to be adequate. Indeed, if the pairing is caused by low-frequency modes, then it requires strong coupling (see below). If, on the other hand, one assumes that the superconductivity is due to weak coupling to high-frequency vibrations Ω_H, one may try to use Eq. (2.46) to account for T_c in the fullerenes. But the fact of the matter is that the BCS expression (2.46) as well as more general treatments making use of the Eliashberg equation are based on the fundamental condition $\Omega \ll E_F$ (the adiabatic Migdal theorem, see Section 2.2.1) which does not hold in the above case. At present, there is no existing theory capable of describing the case $\Omega \gtrsim E_F$.

Fortunately, the major contribution to the pairing in the fullerenes comes from the low-frequency region Ω_L. Let us discuss this question in more detail [120]. As was mentioned above, our goal is to evaluate the values of the coupling constant λ, the Coulomb pseudopotential μ^*, and the characteristic frequency $\tilde{\Omega} = \langle \Omega^2 \rangle^{1/2}$. We start from the general expression for T_c [11c] (see Section 2.3.5):

$$T_c = 0.25\tilde{\Omega}[\exp(2/\lambda_{eff}) - 1]^{-1/2}$$
$$\lambda_{eff} = (\lambda - \mu^*)[1 + 2\mu^* + \lambda\mu^* t(\lambda)]^{-1}, \tag{7.22}$$

Here λ is the electron–phonon coupling constant, $\tilde{\Omega} = \langle \Omega^2 \rangle^{1/2}$, where $\langle \Omega^n \rangle$ is defined in the usual way, so that $\langle \Omega^2 \rangle = 2 \int d\Omega\, \Omega a^2(\Omega)F(\Omega)$. The universal function $t(x)$ is presented in Fig. 2.3b. The isotope coefficient is determined by the relation:

$$\alpha = 0.5(\tilde{\Omega}/T_c)(\partial T_c/\partial \tilde{\Omega}) \tag{7.23}$$

Based on Eqs (7.22), (7.23), (2.55) and the relation $\partial\mu^*/\partial\tilde{\Omega} = (\mu^*)^2\tilde{\Omega}^{-1}$, one can derive the following expression for the isotope coefficient:

$$\alpha = 0.5\left[1 - \frac{\mu^{*2}}{\lambda_{eff}}\frac{1 + 2\lambda + \lambda^2 t(\lambda)}{\lambda - \mu^*}[1 + (4T_c/\tilde{\Omega})^2]\right]. \tag{7.24}$$

T_c and α are experimentally measured quantities. Such measurements can be used in order to determine λ and μ^* for a given value of $\tilde{\Omega}$. In other words, Eqs (7.22) and (7.24) can be treated as a sum of two equations allowing us to evaluate λ and μ^* for given $\tilde{\Omega}$.

In order to complete the analysis, it is useful to consider also the NMR data. The NMR measurements allow one to determine the ratio R of the densities of states for two superconductors, so that $R = v_1/v_2$; see, e.g., [121]. Let us assume that the pairing in fullerenes is due to the intramolecular vibrations (this assumption will be confirmed below). Then the pairing in different fullerenes, like Rb_3C_{60} and K_3C_{60}, is due to the coupling to the same mode Ω. Using the McMillan representation (2.51) for the coupling constant, one can write $R = \lambda_1/\lambda_2$. If we neglect μ^* in a first approximation (since $\mu^* \ll \lambda$), then Eq. (2.61) leads to the following expression for R:

$$R = \ln(1 + \gamma^2 x^2)/\ln(1 + x^2) \tag{7.25}$$

where $\gamma = T_{c1}/T_{c2}$ and $x = 0.25\tilde{\Omega}/T_{c1}$. This equation allows us, in a first approximation, to determine the value of the characteristic phonon frequency $\tilde{\Omega}$. Using Eqs (7.22) and (7.24), one can evaluate the parameters λ_i, μ_i^* ($i = 1, 2$); then one can use the next iteration to increase the accuracy of the calculation.

Based on Eqs (7.22), (7.24), (7.25) and the appropriate experimental data, one can evaluate the major parameters and identify the region of vibrational frequencies which makes a major contribution to the pairing.

Note that if $\Omega \simeq \Omega_{inter} \lesssim 50$ meV, then we obtain from Eqs (7.22) and (7.24) an extremely large value ($\lambda \sim 12$). This value is unreasonably large, and we can focus on the intramolecular modes.

Let us turn to Rb_3C_{60} and K_3C_{60}. According to [121], the value of $R = \lambda_1/\lambda_2 \approx 1.3$–1.4 for M_iC_{60} ($i = 1, 2$; the indices correspond to Rb and K, respectively). Using Eq. (7.25) and the values $R = 1.3$–1.4 [121], $T_{c1} = 28$ K, and $T_{c2} = 19$ K, we obtain $\tilde{\Omega} \approx \Omega_L \approx 250$–300 cm^{-1}.

Let us consider Rb_3C_{60} ($T_c = 28$ K). According to [117], the isotope coefficient $\alpha = 0.37$. As was noted above, Eqs (7.22) and (7.24) allow us to determine λ and μ^* for a given value of $\tilde{\Omega}$.

For example, for the low-frequency intramolecular mode $\tilde{\Omega} \approx \Omega_L \approx 250$ cm^{-1} we find from Eqs (7.22) and (7.24) the following values: $\lambda \approx 2.1$, $\mu^* \approx 0.2$. These values are quite realistic: even some conventional superconductors display coupling constants within this range, e.g., $\lambda = 2.6$ in Am–$Pb_{0.45}Bi_{0.55}$, and $\lambda = 2$ in $Pb_{0.7}Bi_{0.3}$ [57a, 122]; the increase in μ^* with respect to the conventional metals where $\mu^* \cong 0.1$–0.15 reflects the small value of E_F.

It is essential that the above analysis is based on an equation for T_c (Eq. (7.22)) which is applicable at the corresponding values of λ. Otherwise, the calculation would not be self-consistent. The BCS equation ($\lambda \ll 1$),

Eq. (2.46), the McMillan–Dynes expression ($\lambda \lesssim 1.5$), Eq. (2.50), or Eq. (2.60) are all not applicable for $\lambda \approx 2.1$.

Using the values of λ_1 and μ_1^* ($\lambda_1 = 2.1$, $\mu_1^* = 0.2$) obtained above and the relation $\lambda_1/\lambda_2 = R$, one can evaluate (see Eqs (7.22)) the values of λ_2 and μ_2^* for K_2C_{60}; we obtain $\lambda_2 \approx 1.5$–1.55 and $\mu_2^* \approx 0.25$. In addition, we can calculate the isotope coefficient α_2 (see Eq. (7.24)) to be ≈ 0.25.

Now we can use the obtained values of λ_i, μ_i^*, and $\tilde{\Omega}$ in order to calculate T_{ci} and α_i (see Eqs (7.22), (7.24), and the equation, similar to (7.25), with $\mu^* \neq 0$). One can see that these equations are satisfied with high accuracy, which reflects the self-consistency of our approach.

According to the tunneling measurements [123], the ratio $2\varepsilon(0)/T_c \approx 5$ in the superconducting fullerenes; these data support the strong coupling picture just described.

Note the fact that the pairing due to coupling to the intramolecular vibrations makes particularly interesting the possibility of intramolecular doping. In this case one could obtain an isolated C_{60}-based cluster with pair correlation, similar to the situation for finite π-electron systems studied in [124].

Therefore, a theoretical analysis based on Eqs (7.22), (7.24), and (7.25), and on the experimental data [117, 121] leads to the conclusion that the superconducting state in the doped fullerenes is due to strong coupling (λ is equal to ≈ 2.1 and ≈ 1.5 for Rb_3C_{60} and K_3C_{60}, respectively); in addition, $\tilde{\Omega} = \langle \tilde{\Omega}^2 \rangle^{1/2} \approx 250$–$300$ cm^{-1}.

7.8. Future directions

In this chapter we have described our framework for understanding some of the unusual properties of the cuprates. There are still many aspects of their behavior that will require considerable theoretical and experimental effort to be fully understood.

One of the most interesting and least understood aspects of the cuprates is the dynamics of the doping from the magnetic insulating state to the metallic superconducting state and finally to the nonsuperconducting metallic state. How does the Fermi surface form and from what kind of precursor state? How can one understand the complicated evolution of the magnetism, from antiferromagnetism through a spin-fluctuating state to a possibly nonmagnetic state?

There is still no microscopic understanding of the transport and optical properties. There is an unexplained Raman continuum [125] and an anomalous temperature dependence of the relaxation time that is evident in many experiments [75].

Tunneling spectroscopy needs to be more convincing. The data that exists are very suggestive but have too many qualifiers to be the final word. Ultimately, tunneling will provide the necessary information to conclusively confirm the mechanism of superconductivity in the cuprates.

The Fermiology that presently exists is rudimentary. It is extremely important to develop techniques to do detailed Fermiology on many of the more interesting cuprates.

It is still an open question whether the bismuth and thallium cuprates have multiple-gap spectra (see above). More experimental work is needed looking in detail at the spectroscopy of these materials as a function of composition.

Finally, the question of the ultimate limit of T_c in the cuprates is still very open!

There are still many unresolved problems in the physics of high T_c (dynamics of doping, Fermiology of bismuth- and thallium-based cuprates, some aspects of the vortex dynamics, etc.), but nevertheless one can formulate the basic principles of the physics describing the properties of these exotic materials.

REFERENCES

Chapter 1
1. G. Bednorz and K. A. Mueller, *Z. Phys.* **B64**, 189 (1986).
2. (a) *Novel Superconductivity*, S. A. Wolf and V. Z. Kresin, Eds. Plenum Press, New York, 1987. (b) *Physical Properties of High Temperature Superconductors* I, II, III, D. Ginsberg, Ed. World Scientific, Singapore, 1989, 1990, 1992. (c) *High Temperature Superconductivity*, J. Ashkenazi *et al.*, Ed. Plenum Press, New York, 1992. (d) *High Temperature Superconductivity*, World Scientific, Singapore, 1987–1992 series.
3. J. Bardeen, L. Cooper, and J. R. Schrieffer, *Phys. Rev.* **108**, 1175 (1957).
4. G. Eliashberg, *Sov. Phys.—JETP* **13**, 1000 (1961); **16**, 790 (1963).
5. H. K. Onnes, *Leiden Commun.* **124C** (1911).
6. W. Meissner and R. Ochsenfeld, *Naturwissenschaften* **21**, 787 (1933).
7. F. London and H. London, *Proc. Roy. Soc. (London)* **A149**, 71 (1935); see also F. London, *Superfluids*, Vol. 1, Wiley, New York, 1950.
8. V. Ginzburg and L. Landau, *Zh. Eksp. Teor. Fiz.* **20**, 1064 (1950).
9. A. A. Abrikosov, *Zh. Eksp. Teor. Fiz.*, **39**, 1797 (1957).
10. E. Maxwell, *Phys. Rev.* **78**, 477 (1950). C. A. Reynolds *et al.*, *Phys. Rev.* **78**, 487 (1950). H. Fröhlich, *Phys. Rev.* **79**, 845 (1950).
11. (a) I. Giaever, *Phys. Rev. Lett.* **5**, 464 (1960). (b) B. D. Josephson, *Phys. Lett.* **1**, 251 (1962).
12. W. A. Little, *Phys. Rev.* **156**, 396 (1964).
13. D. C. Johnston *et al.*, *Mater. Res. Bull.* **8**, 777 (1973).
14. A. W. Sleight, J. L. Gillson, and P. E. Bierstedt, *Solid State Commun.* **17**, 27 (1975).
15. R. L. Greene, P. M. Grant, and G. B. Street, *Phys. Rev. Lett.* **34**, 89 (1975).
16. D. Jerome *et al.*, *J. Phys. (Paris) Lett.* **41**, L95 (1980).
17. (a) N. Nguyen *et al.*, *J. Solid State Chem.* **39**, 120 (1981). (b) C. Michel and B. Raveau, *Rev. Chem. Miner.* **21**, 407 (1984). (c) L. Er-rakho *et al.*, *J. Solid State Chem.* **37**, 151 (1981).
18. 19. M. K. Wu *et al.*, *Phys. Rev. Lett.* **58**, 908 (1987).
20. H. Maeda *et al.*, *Jpn J. Appl. Phys. Lett.* **27**, 209 (1988).
21. Z. Sheng and A. Hermann, *Nature* **332**, 55 (1988).
22. A. Hebard *et al.*, *Nature* **350**, 600 (1991).

Chapter 2
1. M. Born and R. Oppenheimer, *Ann. Phys.* **84**, 457 (1927).
2. J. Ziman, *Proc. Cambr. Phil. Soc.* **52**, 707 (1955).
3. H. Fröhlich, *Phys. Rev.* **79**, 845 (1950).

4. M. Born and M. Huang, *Dynamic theory of crystal lattice*. Oxford University Press, Oxford, 1954.

5. G. Grimvall, *The Electron–phonon interaction in metals*. North-Holland, Amsterdam, 1981.

6. J. Ziman, *Electrons and phonons*. Clarendon Press, Oxford, 1960.

7. V. Kresin and W. A. Lester, Jr., "The adiabatic methods in the theory of many-body systems," in *Mathematics analysis of physical systems*, p. 247, R. Mickens, Ed. Van Nostrand, New York, 1985.

8. A. Migdal, *Qualitative methods in quantum theory*. W. Benjamin, New York, 1977.

9. B. Geilikman, *J. Low Temp. Phys.* **4**, 189 (1971).

10. Be. Geilikman, *Sov. Phys.—USP* **18**, 190 (1975).

11. J. Bardeen, L. Cooper, and J. Schrieffer, *Phys. Rev.* **108**, 175 (1957).

12. P. Allen and R. Dynes, *Phys. Rev.* **312**, 905 (1975).

13. E. Wolf, *Principles of electron tunneling spectroscopy*. Oxford Press, New York, 1985.

14. G. Eliashberg, *Sov. Phys.—JETP* **11**, 696 (1960); **12**, 1000 (1961).

15. A. Migdal, *Sov. Phys.—JETP* **7**, 996 (1958).

16. A. Abrikosov, L. Gor'kov, and I. Dzyaloshinski, *Methods of quantum field theory in statistical physics*. Dover, New York, 1963.

17. D. Scalapino, in *Superconductivity*, R. Parks, Ed., p. 449. Marcel Dekker, New York, 1969.

18. E. Lifshitz and L. Pitaevskii, *Statistical Physics II*. Pergamon, New York, 1988.

19. L. Cooper, *Phys. Rev.* **104**, 1189 (1956).

20. L. Gor'kov, *Sov. Phys.—JETP* **7**, 505 (1958).

21. S. Louie and M. L. Cohen, *Solid State Commun.* **22**, 1 (1977).

22. V. Kresin, *J. Low Temp. Phys.* **5**, 505 (1971).

23. B. Geilikman, *J. Low Temp. Phys.* **4**, 181 (1971).

24. W. McMillman, *Phys. Rev.* **167**, 331 (1968).

25. R. Dynes, *Solid State Commun.* **10**, 615 (1972).

26. B. Geilikman, V. Kresin, and N. Masharov, *J. Low Temp. Phys.* **18**, 241 (1975).

27. B. Geilikman and N. Masharov, *Phys. Stat. Sol.* **41**, K31 (1970).

28. N. Bogoluybov, N. Tolmachev, and D. Shirkov, *A new method in the theory of superconductivity*. Cons. Bureau, New York, 1959.

29. P. Morel and P. Anderson, *Phys. Rev.* **125**, 1263 (1962).

30. A. Abrikosov and I. Khalatnikov, *Adv. Phys.* **8**, 45 (1959).

31. R. Schneider, J. Geerk, and H. Rietchel, *Europhys. Lett.* **4**, 845 (1987).

32. V. Kresin, H. Gutfreund, and W. A. Little, *Solid State Commun.* **51**, 339 (1984).

33. C. Owen and D. Scalapino, *Physica* **55**, 691 (1971).

34. V. Kresin, *Bull. Am. Phys. Soc.* **32**, 796 (1976); *Phys. Lett A* **122**, 434 (1987).

35. L. Bourne *et al.*, *Phys. Rev. B* **36**, 3990 (1987).

36. B. Geilikman and V. Kresin, *Sov. Phys.—Solid State* **7**, 2659 (1966); *Phys. Lett.* **40A**, 123 (1972).

37. N. Zubarev, *Sov. Phys.—USP* **3**, 320 (1960).

38. D. Douglass, Jr. and L. Falicov, in *Progress in low temperature physics*, Vol. IV, C. Gorter, Ed., p. 97. North-Holland, Amsterdam, 1968.

39. N. Zavaritskii, E. Itskevich, and A. Voronovskii, *Sov. Phys.—JETP* **33**, 762 (1971).

40. J. Carbotte, *Rev. Mod. Phys.* **62**, 1027 (1990).

41. V. Kresin, *Solid State Commun.* **63**, 725 (1987).

42. J. Carbotte, in *Novel superconductivity*, S. Wolf and V. Kresin, Eds., p. 73. Plenum Press, New York, 1987.
43. V. Kresin and A. Parchomenko, *Sov. Phys.—Solid State* **16**, 2180 (1975).
44. G. Eliashberg, *Sov. Phys.—JETP* **16**, 780 (1963).
45. B. Geilikman and V. Kresin, *Sov. Phys.—Solid State* **9**, 2453 (1968).
46. D. Scalapino, J. Schrieffer, and J. Wilkins, *Phys. Rev.* **148**, 263 (1966).
47. G. Grimvall, *Phys. kondens. Materie* **9**, 283 (1969).
48. Y. Kresin and R. Zaitsev, *Sov. Phys.—JETP* **47**, 983 (1978).
49. Y. Browman and Yu. Kagan, *Sov. Phys.—JETP* **25**, 365 (1967).
50. R. Blinc and B. Zeks, in *Soft modes in ferroelectrics and anti-ferroelectrics*, Series of Monographs on Selected Topics in Solid State Physics, E. P. Wohlfarth, Ed. North-Holland, Amsterdam and Oxford, 1974.
51. S. M. Shapiro *et al.*, *Phys. Rev. B* **6**, 4332 (1972).
52. J. K. Kjems *et al.*, *Phys. Rev. B* **8**, 1119 (1973).
53. P. Boeni *et al.*, *Phys. Rev. B* **38**, 185 (1988).
54. G. Bednorz and K. A. Mueller, *Z. Phys. B* **64**, 189 (1986).
55. J. D. Axe *et al.*, *Phys. Rev. Lett.* **62**, 2751 (1989).
56. R. J. Moodenbaugh *et al.*, *Phys. Rev. B* **38**, 4596 (1988).
57. T. Suzuki and T. Fujita, *J. Phys. Soc. Japan* **58**, 1883 (1989).
58. M. K. Crawford *et al.*, *Phys. Rev. B* **41**, 282 (1990).
59. H. Morawitz, *Phys. Rev. Lett.* **34**, 1567 (1975).
60. R. E. Cohen *et al.*, *Phys. Rev. Lett.* **60**, 817 (1988); R. E. Cohen, W. E. Pickett, and K. Krakauer, *Phys. Rev. Lett.* **62**, 831 (1989); H. Krakauer, *Phys. Rev. Lett.* **64**, 2575 (1990); see also W. E. Pickett, *Rev. Mod. Phys.* **61**, 433 (1989).
61. B. H. Toby *et al.*, *Phys. Rev. Lett.* **64**, 2414 (1990).
62. J. Mustre de Leon *et al.*, *Phys. Rev. Lett.* **65**, 1675 (1990).
63. R. P. Sharma *et al.*, *Phys. Rev. Lett.* **62**, 2869 (1990).
64. W. Reichardt *et al.*, *Physica C* **163–164**, 464 (1989).
65. H. Rietschel, L. Pintschovius, and W. Reichardt, *Physica C* **162–164**, 1705 (1989).
66. (a) H. Morawitz and V. Z. Kresin, *Bull. APS* **29**, 2234 (1989). (b) H. Morawitz and V. Z. Kresin, in *The electronic structure of high T_c superconductors*, J. Fink, H. Kuzmany, and G. Mehring, Eds. Springer Verlag, Kirchberg, Austria (1990).
67. H. Bilz, G. Benedek, and A. Bussmann-Holder, *Phys. Rev. B* **35**, 4840 (1987).
68. A. Bussmann-Holder *et al.*, *Z. Phys. B.—Cond. Matter* **79**, 445 (1990).
69. A. Bussmann-Holder, H. Bilz, and G. Benedek, *Phys. Rev. B* **39**, 9214 (1989).
70. A. Bussmann-Holder, A. Simon, and H. Buetter, *Phys. Rev. B* **39**, 207 (1989).
71. N. S. Gillis and T. R. Koehler *Phys. Rev. B* **5**, 1924 (1972).
72. K. L. Ngai, *Phys. Rev. Lett.* **52**, 215 (1975).
73. J. F. Schooley *et al.*, *Phys. Rev. Lett.* **14**, 305 (1965).
74. H. Morawitz and V. Z. Kresin, *Bull. APS* **36**, 613 (1991).
75. D. J. Scalapino, Y. Wada, and J. C. Swihart, *Phys. Rev. Lett.* **14**, 102 (1965).
76. J. M. Rowell and W. L. McMillan, *Phys. Rev. Lett.* **14**, 108 (1965).
77. A. W. Sleight *et al.*, *Solid State Commun.* **17**, 27 (1976).
78. J. C. K. Hui and P. W. Allen, *J. Phys. F* **4**, L42 (1974).
79. J. R. Hardy and J. W. Flocken, *Phys. Rev. Lett.* **60**, 2191 (1988).
80. N. M. Plakida, V. L. Aksenov, and S. L. Drechsler, *Europhys. Lett.* **4**, 1309 (1987).
81. H. Morawitz and V. Z. Kresin (unpublished).

82. G. Bergmann and D. Rainer, *Z. Phys.* **263**, 59 (1973).
83. A. E. Karakozov and E. G. Maksimov, *Zh. Eksp. Teor. Fiz.* **74**, 681 (1978) (*Sov. Phys.—JETP* **47**, 358 (1978)).
84. D. J. Scalapino, R. T. Scalettar, and N. Bickers, Ref. 42, p. 475.
85. H. B. Schuettler and A. J. Fedro, *Phys. Rev. B* **39**, 2484 (1989).
86. M. R. Schafroth, *Phys. Rev.* **100**, 463 (1955).
87. N. F. Mott, *Nature* **327**, 185 (1987); *Phil. Mag. Lett.* **63**, 319 (1991).
88. (a) T. M. Rice, *Nature* **332**, 780 (1988). (b) A. Broyles, E. Teller, and B. Wilson, *J. Superconductivity* **3**, 161 (1990).
89. V. L. Vinetskii, *Zh. Eksp. Teor. Fiz.* **40**, 1459 (1961) (*Sov. Phys.—JETP* **13**, 1023 (1961)).
90. D. M. Eagles, *Phys. Rev.* **186**, 456 (1969).
91. T. Holstein, *Ann. Phys. (NY)* **8**, 343 (1959).
92. I. G. Lang and Yu. A. Firsov, *Zh. Eksp. Teor. Fiz.* **43**, 1843 (1962 (*Sov. Phys.—JETP* **16**, 1301 (1963)).
93. P. W. Anderson, *Phys. Rev. Lett.* **34**, 953 (1975).
94. P. Pincus, P. Chaikin, and C. F. Coll III, *Solid State Commun.* **12**, 1265 (1973).
95. Y. Nagaoka, *Prog. Theor. Phys.* **52**, 1716 (1974).
96. Ph. Nozieres and S. Schmitt-Rink, *J. Low Temp. Phys.* **59**, 195 (1985).
97. A. Alexandrov and J. Ranninger, *Phys. Rev. B* **23**, 1796 (1981).
98. S. Lakkis *et al.*, *Phys. Rev. B* **14**, 1429 (1976).
99. B. K. Chakraverty, M. T. Sienko, and J. Bonnerot, *Phys. Rev. B* **17**, 3781 (1978).
100. A. Alexandrov, J. Ranninger, and S. Robaszkiewicz, *Phys. Rev. B* **33**, 4526 (1986).
101. D. Emin, *Phys. Rev. Lett.* **62**, 1544 (1989).
102. L. J. DeJongh, *Physica C* **161**, 631 (1989).
103. R. Micnas, J. Ranninger, and S. Robaczkiewicz, *Rev. Mod. Phys.* **62**, 113 (1990).
104. B. Geilikman, *Sov. Phys.—Solid State* **18**, 54 (1976).
105. A. A. Maradudin, E. W. Montroll, and G. H. Weiss, *Theory of lattice dynamics in the harmonic approximation*, Suppl. 3 to Solid State Physics. Academic Press, New York, 1963.
106. (a) B. M. Klein, E. N. Economou, and D. A. Papaconstantoupoulos, *Phys. Rev. Lett.* **39**, 574 (1977). (b) D. A. Papaconstantopoulos *et al.*, *Phys. Rev. B* **17**, 141 (1978). (c) B. M. Klein and R. E. Cohen, *Phys. Rev. B* **45**, 12405 (1992).

Chapter 3

1. G. M. Eliashberg, *Sov. Phys.—JETP* **11**, 696 (1960).
2. W. L. McMillan and J. M. Rowell, *Phys. Rev. Lett.* **14**, 108 (1965).
3. W. L. McMillan and J. M. Rowell, *Superconductivity*, p. 561, R. D. Parks, Ed. Marcel Dekker, New York, 1969.
4. E. L. Wolf, *Principles of electron tunneling spectroscopy*. Oxford University Press, Oxford, 1985.
5. J. Nicol, S. Shapiro, and P. H. Smith, *Phys. Rev. Lett.* **5**, 461 (1960).
6. S. Bermoh, *Tech. Rept. 1*, University of Illinois, Urbana, National Science Foundation Grant NSF GP1100, 1964.
7. I. Giaever and K. Megerle, *Phys. Rev.* **122**, 1101 (1961).
8. D. J. Scalapino, J. M. Schrieffer, and J. M. Wilkins, *Phys. Rev.* **148**, 263 (1966).
9. R. Schneider, J. Geerk, and H. Reitschel, *Europhys. Lett.* **4**, 845 (1987).
10. V. Z. Kresin, *Phys. Rev.* **6**, 122 (1987).

11. V. Z. Kresin, *Phys. Rev. B* **30**, 450 (1984).
12. K. Kihlstrom, P. V. Hovda, V. Z. Kresin, and S. A. Wolf, *Phys. Rev. B* **38**, 4588 (1988).
13. D. C. Mattis and J. Bardeen, *Phys. Rev.* **111**, 412 (1958).
14. A. A. Abrikosov, L. P. Gor'kov, and I. M. Khalatnikov, *Sov. Phys.—JETP* **8**, 182 (1959).
15. D. M. Ginsberg and L. C. Hebel, Ref. 3, p. 193.
16. R. E. Glover III and M. Tinkham, *Phys. Rev.* **108**, 243 (1957).
17. L. H. Palmer, PhD Thesis, University of California, Berkeley, 1966 (unpublished); Ref. 3, p. 209.
18. J. Bardeen, L. N. Cooper, and J. R. Schrieffer, *Phys. Rev.* **108**, 1175 (1957).
19. V. Z. Kresin, *Sov. Phys.—JETP* **9**, 1385 (1959).
20. R. W. Morse and H. V. Bohm, *Phys. Rev.* **108**, 1094 (1957).
21. L. C. Hebel and C. P. Slichter, *Phys. Rev.* **107**, 901 (1957); A. G. Redfield, *Phys. Rev. Lett.* **3**, 85 (1959); Y. Masuda and A. G. Redfield, *Phys. Rev.* **125**, 159 (1962).
22. (a) Eliashberg, unpublished. (b) P. B. Allen and D. Rainer, *Nature* **349**, 396 (1991).
23. K. Yoshida, *Phys. Rev.* **106**, 893 (1957).

Chapter 4
1. W. A. Little, *Phys. Rev.* **134**, A1416 (1964).
2. R. Greene, G. Street, and L. Suter, *Phys. Rev. Lett.* **34**, 577 (1975).
3. W. Gill *et al.*, *Phys. Rev. Lett.* **35**, 1732 (1975).
4. D. Jerome *et al.*, *Phys. Lett.* **41**, L-95 (1980).
5. J. Ferraro and J. Williams, *Introduction to synthetic electrical conductors.* Academic Press, New York, 1987.
6. T. Ishiguro and K. Yamaji, *Organic superconductors.* Springer-Verlag, Berlin, 1990.
7. V. Kresin and W. A. Little, Eds., *Organic superconductivity.* Plenum Press, New York, 1990.
8. H. Gutfreund and W. A. Little, in *Highly conducting one-dimensional solids*, J. Devreese, R. Evrard, and Va. van Doren, Eds., p. 305. Plenum Press, New York, 1979.
9. H. Gutfreund and W. A. Little, *Rice University Studies* **66**, 1 (1980).
10. W. A. Little, *J. de Phys. Coll. C* **3**, 819 (1983).
11. R. Ferrell, *Phys. Rev. Lett.* **13**, 330 (1964).
12. T. M. Rice, *Phys. Rev.* **140**, A889 (1965).
13. P. Hohenberg, *Phys. Rev.* **158**, 383 (1967).
14. I. Dzyaloshinskii and E. Katz, *Sov. Phys.—JETP* **55**, 338 (1968).
15. Yu. Bychkov, L. Gor'kov, and I. Dzyaloshinskii, *Sov. Phys.—JETP* **23**, 489 (1966).
16. I. Dzyaloshinskii and A. Larkin, *Sov. Phys.—JETP* **34**, 422 (1972).
17. V. Ginzburg, *Sov. Phys.—JETP* **20**, 1549 (1965).
18. V. Ginzburg and D. Kirvzhnits, Eds., *High temperature superconductivity.* Cons. Bureau, New York, 1982.
19. D. Allender, J. Bray, and J. Bardeen, *Phys. Rev. B* **7**, 1020 (1973).
20. W. A. Little, in *Novel superconductivity*, S. Wolf and V. Kresin, Eds., p. 341. Plenum Press, New York, 1987.
21. J. Yu, A. Freeman, and S. Massidda, Ref. 20, p. 367.
22. W. Hsu and R. Kasowski, Ref. 20, p. 373.

23. C. Varma, S. Schmitt-Rink, and E. Abrahams, Ref. 20, p. 355.
24. P. Littlewood, Lectures at 9th Jerusalem Winter School for Theoretical Physics, 1992.
25. B. Geilikman, *Sov. Phys.—JETP* **48**, 1194 (1965).
26. B. Geilikman, *J. Low Temp. Phys.* **4**, 189 (1971).
27. B. Geilikman, *Sov. Phys.—JETP* **8**, 2032 (1966); **16**, 17 (1973).
28. H. Gutfreund, Ref. 20, p. 465.
29. V. Kresin, *Phys. Lett.* **49A**, 117 (1974).
30. B. Geilikman and V. Z. Kresin. *Sov. Phys.—Semicond.* **2**, 639 (1968).
31. P. Anderson, *Phys. Rev. Lett.* **34**, 953 (1975).
32. C. Ting, D. Talwar, and K. Ngai, *Phys. Rev. Lett.* **45**, 1213 (1980).
33. J. Hirsch and J. Scalapino, *Phys. Rev. B* **32**, 5639 (1985).
34. H. Schuttler, M. Jarrell, and D. Scalapino, *Phys. Rev. Lett.* **58**, 1147 (1987); *Phys. Rev. B* **39**, 6501 (1989).
35. H. Schuttler, M. Jarrell, and D. Scalapino, Ref. 20, p. 481.
36. Y. Bar-Yam, *Phys. Rev.* **43**, 359, 2601 (1991).
37. Y. Bar-Yam, in *High temperature superconductivity*, J. Askenazi *et al.*, Eds., p. 561. Plenum Press, New York, 1991.
38. J. Goldman, *Zh. Eksp. Teor. Fiz.* **17**, 681 (1947).
39. D. Bohm and D. Pines, *Phys. Rev.* **82**, 625 (1951).
40. D. Pines, *Can. J. Phys.* **34**, 1379 (1956).
41. (a) H. Fröhlich, *J. Phys. C* **1**, 544 (1968). (b) J. Garland, *Phys. Rev. Lett.* **11**, 111, 114 (1963). (c) B. Geilikman, *Sov. Phys.—USP* **8**, 2032 (1966). (d) E. Pashitskii, *Sov. Phys.—JETP* **28**, 1267 (1968). (e) A. Rothwarf, *Phys. Rev. B* **2**, 3560 (1970). (f) J. Ihm, M. L. Cohen, and S. Tuan, *Phys. Rev. B* **23**, 3258 (1981).
42. (a) J. Ruvalds, *Adv. Phys.* **30**, 677 (1981). (b) J. Ruvalds, *Phys. Rev. B* **35**, 8869 (1987).
43. F. Stern, *Phys. Rev. Lett.* **18**, 5646 (1967).
44. Y. Takada, *J. Phys. Soc. Jpn*, **45**, 786 (1978); **49**, 1713 (1980).
45. V. Kresin and H. Morawitz, *Phys. Rev. B* **37**, 7854 (1988); *J. Supercond.* **1**, 108 (1988).
46. J. Ashkenazi, D. Vacaru, and C. Kuper, Ref. 37, p. 569.

Chapter 5

1. N. F. Berk and J. R. Schrieffer, *Phys. Rev. Lett.* **17**, 433 (1966).
2. S. Doniach and S. Engelsberg, *Phys. Rev. Lett.* **17**, 750 (1966).
3. P. W. Anderson and W. F. Brinkman, *Phys. Rev. Lett.* **30**, 1108 (1973).
4. D. J. Scalapino, E. Loh, and J. E. Hirsch, *Phys. Rev. B* **34**, 8190 (1986).
5. K. Miyake, S. Schmitt-Rink, and C. M. Varma, *Phys. Rev. B* **34**, 6554 (1986).
6. D. J. Scalapino, E. Loh, and J. E. Hirsch, *Phys. Rev. B* **35**, 6694 (1987).
7. J. G. Bednorz and K. A. Mueller, *Z. Phys. B* **64**, 189 (1986).
8. J. R. Schrieffer, X. G. Wen, and S. C. Zhang, *Phys. Rev. B* **39**, 11663 (1989).
9. V. J. Emery, *Phys. Rev. Lett.* **58**, 3759 (1987).
10. F. C. Zhang and T. M. Rice, *Phys. Rev. B* **57**, 3759 (1988).
11. G. Kotliar, P. A. Lee, and N. Read, *Physica C* **153–155**, 528 (1988).
12. G. Kotliar and A. E. Ruckenstein, *Phys. Rev. Lett.* **57**, 1362 (1986).
13. R. T. Scalettar *et al.*, *Phys. Rev. Lett.* **62**, 1407 (1989).
14. B. I. Shraiman and E. D. Siggia, *Phys. Rev. Lett.* **61**, 467 (1988).
15. B. I. Shraiman and E. D. Siggia, *Phys. Rev. Lett.* **62**, 1564 (1989).

16. P. W. Anderson, *Science* **235**, 1196 (1987). P. W. Anderson, G. Baskaran, Z. Zou, and T. Hsu, *Phys. Rev. Lett.* **58**, 2790 (1987).
17. S. Kivelson, D. Rokhsar, and J. Sethna, *Phys. Rev. B* **35**, 8865 (1987).
18. V. Kalmeyer and R. B. Laughlin, *Phys. Rev. Lett.* **59**, 2095 (1987).
19. R. B. Laughlin, *Science* **242**, 525 (1988).
20. N. F. Mott, *Rev. Mod. Phys.* **40**, 677 (1968).
21. P. W. Anderson, *Solid State Phys.* **14**, 99 (1963).
22. J. Hubbard, *Proc. Roy. Soc. (London) Ser. A* **276**, 238 (1963); **281**, 401 (1964).
23. S. Chakravarty, D. Nelson, and B. I. Halperin, *Phys. Rev. Lett.* **60**, 1057 (1988).
24. L. F. Matheiss, *Phys. Rev. Lett.* **58**, 1028 (1987).
25. J. Yu, A. J. Freeman, and J. H. Xu, *Phys. Rev. Lett.* **58**, 1035 (1987).
26. R. E. Cohen, W. E. Pickett, L. L. Boyer, and H. Krakauer, *Phys. Rev. Lett.* **60**, 817 (1987). W. E. Pickett, *Rev. Mod. Phys.* **61**, 433 (1989).
27. J. B. Torrance and R. M. Metzger, *Phys. Rev. Lett.* **63**, 1515 (1989).
28. J. Kondo, Y. Asai, and S. Nagai, *J. Phys. Soc. Japan* **57**, 4334 (1988).
29. L. F. Feiner and D. M. deLeeuw, *Solid State Commun.* **70**, 1165 (1989).
30. A. K. McMahan, J. F. Annett, and R. M. Martin, *Phys. Rev. B* **42**, 6268 (1990).
31. M. S. Hybertsen, M. Schlueter, and N. E. Christensen, *Phys. Rev. B* **39**, 9028 (1989).
32. W. Weber, *Z. Phys. B* **70**, 323 (1988).
33. J. Zaanen, A. M. Oles, and L. F. Feiner, in *Dynamics of magnetic fluctuations in high T_c superconductors*, Vol. 246 of NATO Advanced Study Institute, Series B: Physics, G. Reiter, P. Horsch, and G. Psaltakis, Eds. Plenum Press, New York, 1991.
34. D. P. Clougherty, K. Johnson, and M. E. Henry, *Physica C* **162–164**, 1475 (1989).
35. L. Krusin-Elbaum *et al.*, *Phys. Rev. Lett.* **62**, 217 (1989).
36. C. G. Olsen *et al.*, *Phys. Rev. B* **42**, 381 (1990).
37. G. Shirane, R. J. Birgeneau *et al.*, *Phys. Rev. Lett.* **63**, 330 (1989).
38. A. Kampf and J. R. Schrieffer, *Phys. Rev. B* **41**, 11663 (1990).
39. A. Kampf and J. R. Schrieffer, *Phys. Rev. B* **42**, 6399 (1990).
40. W. Brenig *et al.*, *Phys. Rev. B* **43**, 258 (1991).
41. J. M. Tranquada *et al.*, *Phys. Rev. B* **35**, 7187 (1987).
42. J. M. Tranquada *et al.*, *Phys. Rev. B* **36**, 5263 (1988).
43. N. Nuecker, H. Romberg, X. X. Xi, J. Fink, D. Hahn, T. Zetterer, H. H. Otto, and K. F. Renk, *Phys. Rev. B* **39**, 6619 (1989).
44. H. Romberg *et al.*, *Phys. Rev. B* **41**, 2609 (1990).
45. D. J. Scalapino, in *Physics of highly correlated electron systems*, K. S. Bedell, D. Coffey, D. Pines, and J. R. Schrieffer, Eds. Addison-Wesley, New York, 1990.
46. R. L. Stratonovich, *Dokl. Akad. Nauk SSSR* **115**, 1097 (1957).
47. J. Hubbard, *Phys. Rev. Lett.* **3**, 77 (1959).
48. S. Q. Wang, W. E. Evanson, and J. R. Schrieffer, *Phys. Rev.* **23**, 92 (1969).
49. J. D. Reger and A. P. Young, *Phys. Rev. B* **37**, 5978 (1988).
50. S. E. Barnes, *J. Phys. F* **6**, 1375 (1976).
51. N. Read and D. Newns, *J. Phys. C* **16**, 3273 (1983). N. Read, *J. Phys. C* **18**, 2651 (1985).
52. P. Coleman, *Phys. Rev. B* **29**, 3035 (1984).
53. A. J. Millis and P. A. Lee, *Phys. Rev. B* **35**, 3394 (1987).
54. A. Auerbach and K. Levin, *Phys. Rev. Lett.* **57**, 877 (1986).
55. J. Zaanen, G. Sawatzky, and J. Allen, *Phys. Rev. Lett.* **55**, 418 (1985).

56. M. Grilli, G. Kotliar, and A. J. Millis, *Phys. Rev. B* **42**, 329 (1990).
57. M. Grilli, C. Castellani, and C. DiCastro, *Phys. Rev. B* **42**, 6233 (1990).
58. C. Di Castro, L. F. Feiner, and M. Grilli, *Phys. Rev. Lett.* **66**, 3209 (1991).
59. L. F. Feiner, M. Grilli, and C. Di Castro, *Phys. Rev. B* **45**, 1047 (1992).
60. P. W. Anderson, *Mater. Res. Bull.* **8**, 153 (1973).
61. P. W. Anderson, Proceedings of the Enrico Fermi International School of Physics, *Frontiers and borderlines in many particle physics*, J. R. Schrieffer and R. A. Broglia, Eds. North-Holland, Amsterdam, 1987.
62. P. W. Anderson, Proceedings of the Common Trends in Particle and Condensed Matter Physics, *Physics Reports*, **184**, 2–4 (1988).
63. P. W. Anderson, unpublished.
64. F. D. M. Haldane, *Phys. Lett.* **81A**, 153 (1981).
65. N. P. Ong *et al.*, *Phys. Rev. Lett.* **67**, 2088 (1991).
66. P. W. Anderson, *Phys. Rev. Lett.* **67**, 2092 (1991).
67. F. Wilczek, *Phys. Rev. Lett.* **49**, 957 (1982).
68. R. B. Laughlin, *Phys. Rev. Lett.* **50**, 1395 (1983).
69. B. Halperin, *Phys. Rev. Lett.* **52**, 1583 (1984).
70. D. Arovas, R. Schrieffer, and F. Wilczek, *Phys. Rev. Lett.* **53**, 722 (1984).
71. 72. R. B. Laughlin, in *The quantum Hall effect*, p. 233, R. E. Prange and S. M. Girvin, Eds. Springer-Verlag, New York, 1986.
73. X. G. Wen, F. Wilczek, and A. Zee, *Phys. Rev. B* **39**, 11413 (1989).
74. R. B. Laughlin, *Phys. Rev. Lett.* **60**, 2677 (1988).
75. I. Dzyaloshinskii, A. Polyakov, and P. Wiegman, *Phys. Rev. Lett.* **127**, 112 (1988).
76. R. Kiefl *et al.*, *Phys. Rev. Lett.* **64**, 2082 (1990).
77. B. Halperin, *Phys. Rev. B* **49**, 1111 (1991).
78. S. Spielman *et al.*, *Phys. Rev. Lett.* **65**, 123 (1990).
79. K. S. Bedell, D. Coffey, D. Pines, and J. R. Schrieffer, Eds., *Physics of highly correlated electron systems*. Addison-Wesley, New York, 1990.
80. G. Baskaran, A. E. Ruchenstein, E. Tosatti, and Y. Lu, Eds., Proceedings of the Adriatico Research Conference and Miniworkshop—Strongly Correlated Electron Systems II. *Progress in high-temperature superconductivity*, Vol. 29. World Scientific, Singapore, 1991.
81. J. Mueller and J. L. Olsen, *Proceedings of materials and mechanisms of high temperature superconductivity I*. North-Holland, Amsterdam, 1988.
82. R. N. Shelton, W. Harrison, and N. E. Philipps, Eds., *Proceedings of materials and mechanisms of high temperature superconductivity II*. North-Holland, Amsterdam, 1989.
83. M. Tachiki, Y. Muto, and Y. Syono, Eds., *Proceedings of materials and mechanisms of high temperature superconductivity III*. North-Holland, Amsterdam, 1991.

Chapter 6

1. H. Suhl, B. Mattias, and L. Walker, *Phys. Rev. Lett.* **3**, 552 (1958).
2. V. Moskalenko, *Fizika Metal Metalloved* **8**, 503 (1959).
3. B. Geilikman, R. Zaitsev, and V. Kresin, *Sov. Phys.—Solid State* **9**, 642 (1967). B. Geilikman and V. Kresin, in *Kinetic and non-stationary phenomena in superconductors*. Wiley, New York, 1974.
4. V. Kresin, *J. Low Temp. Phys.* **11**, 519 (1973).
5. V. Kresin and S. Wolf, *Physica C* **169**, 476 (1990).

6. V. Pokrovsky, *Sov. Phys.—JETP* **13**, 447 (1961).
7. P. Anderson, *J. Phys. Chem. Sol.* **11**, 26 (1959).
8. G. Binnig *et al.*, *Phys. Rev. Lett.* **45**, 1352 (1980).
9. H. Meissner, *Phys. Rev.* **117**, 672 (1960).
10. R. Simon and P. Chaikin, *Phys. Rev. B* **23**, 4463 (1981); **30**, 5552 (1984).
11. J. Clarke, *Proc. Roy. Soc. (London) Ser. A* **308**, 447 (1969).
12. A. Mota, P. Visani, and A. Pollini, *J. Low Temp. Phys.* **76**, 465 (1989).
13. P. G. de Gennes, *Rev. Mod. Phys.* **36**, 225 (1964). G. Deutscher and P. G. de Gennes, in *Superconductivity*, R. Paris, Ed., p. 1005. Marcel Dekker, New York, 1969.
14. W. McMillan, *Phys. Rev.* **175**, 537 (1968).
15. V. Kresin, *Phys. Rev. B* **25**, 157 (1982).
16. J. Bardeen, *Phys. Rev. Lett.* **6**, 57 (1961); **9**, 147 (1962).
17. W. Harrison, *Phys. Rev.* **123**, 85 (1961).
18. L. Cooper, *Phys. Rev. Lett.* **6**, 689 (1961).
19. V. Kresin, S. Wolf, and G. Deutscher, *Physica C* **191**, 9 (1992). V. Kresin and S. Wolf, *Physica C* **198**, 328 (1992).
20. V. Kresin and S. Wolf *Phys. Rev. B* **46**, 6458 (1992).
21. M. H. Cohen, L. Falicov, and J. Phillips, *Phys. Rev. Lett.* **8**, 316 (1962).
22. M. Belogolovskii, Yu. Ivanchencko, and Yu. Medvedev, *Sov. Phys.—JETP Lett.* **21**, 332 (1975); *Sov. Pnys.—Solid State* **17**, 1937 (1976).
23. E. Wolf, *Principles of electron tunneling spectroscopy*, p. 285. Oxford University Press, New York.
24. V. Kresin, *J. Low Temp. Phys.* **5**, 505 (1971).
25. C. Owen and D. Scalapino, *Physica* **55**, 691 (1971).
26. A. Abrikosov and L. Gor'kov, *Sov. Phys.—JETP* **12**, 1243 (1961).
27. A. Kaiser and M. Zuckerman, *Phys. Rev. B* **1**, 229 (1970).
28. G. Deutcher *et al.*, *Phys. Rev. B* **8**, 5055 (1973).

Chapter 7

1. G. Bednorz and K. A. Mueller, *Z. Phys. B* **64**, 189 (1986).
2. (a) B. Raveau *et al.*, *Crystal chemistry at high T_c superconducting high T_c oxides*. Springer-Verlag, Berlin, 1991. (b) C. Michel and B. Raveau, *Rev. Chem. Miner.* **21**, 407 (1984). (c) R. Retoux *et al.*, *Phys. Rev. B* **41**, 193 (1990).
3. M. Wu *et al.*, *Phys. Rev. Lett.* **58**, 908 (1987).
4. (a) H. Maeda *et al.*, *Jpn J. Appl. Phys. Lett.* **27**, 209 (1988). (b) M. Azuma *et al.*, *Nature* **356**, 775 (1992).
5. Z. Sheng and A. Hermann, *Nature* **332**, 55 (1988).
6. V. Kresin and S. Wolf, (a) *Solid State Commun.* **63**, 1141 (1987); *J. Superconductivity*, **1**, 327 (1988). (b) *Phys. Rev. B* **41**, 4178 (1990). (c) *Physica C* **169**, 476 (1990). (d) *Physica C* **198**, 328 (1992). (e) *Phys. Rev. B* **46**, 6458 (1992).
7. V. Kresin, S. Wolf, and G. Deutscher, (a) *J. Supercond.* **1**, 327 (1988); (b) *Physica C* **191**, 9 (1992).
8. S. Wolf and V. Kresin, in *Advances in superconductivity II*, T. Ishiguro and K. Kajimura, Eds., p. 447. Springer-Verlag, Tokyo, 1990.
9. V. Kresin and H. Morawitz, (a) in *Novel superconductivity*, S. Wolf and V. Kresin, Eds., p. 445, 1987. Plenum Press, New York. (b) *Phys. Rev. B* **37**, 7854 (1988). (c) *J. Supercond.* **1**, 89 (1988); *Physica C* **153–155**, 1327 (1988). (d) *J. Supercond.* **3**, 227 (1990). (e) *Solid State Commun.* **74**, 1203 (1990). (f) *Phys. Lett.* **145**, 368 (1990). (g) *Phys. Rev. B* **43**, 2691 (1991).

10. (a) H. Morawitz and V. Kresin, in *Electronic properties of high T_c and related compounds*, H. Kuzmany, M. Mehring, and J. Fink, Eds., p. 376. Springer-Verlag, Berlin, 1990. (b) H. Morawitz *et al.*, *Z. Phys.* (in press).

11. V. Kresin, (a) *Phys. Rev. B* **35**, 8716 (1987). (b) *Solid State Commun.* **51**, 339 (1987). (c) *Phys. Lett. A* **122**, 434 (1987). (d) Ref. 9a, p. 309. (e) *J. Supercond.* **3**, 177 (1990).

12. J. Cohn *et al.*, (a) *Phys. Rev. Lett.* **66**, 1098 (1991). (b) *Phys. Rev. B* **45**, 13140, 13144 (1992). (c) in *High temperature superconductivity*, J. Askenazi *et al.*, Eds., p. 235. Plenum Press, New York, 1991.

13. M. Reeves *et al.*, *Phys. Rev. B* **47**, 6065 (1993).

14. M. Osofsky *et al.*, *Phys. Rev. B* **45**, 4916 (1992).

15. V. Kresin, H. Morawitz, and S. Wolf, in *High temperature superconductivity*, J. Askenazi *et al.*, Eds., p. 275. Plenum Press, New York, 1991.

16. A. Antonello, V. Kresin, and S. Wolf, *J. Supercond.* **5**, 339 (1992).

17. N. Ashcroft and N. Mervin, *Solid state physics*. Holt, Rinehart, and Winston, New York, 1976.

18. A. Cracknell and K. Wong, *The Fermi surface*. Clarendon Press, Oxford, 1973.

19. L. Landau, *Sov. Phys.—JETP* **3**, 920 (1956); **5**, 101 (1957).

20. E. Lifshitz and L. Pitaevskii, *Statistical physics*, p. II, Pergamon, New York, 1988.

21. (a) N. Phillips *et al.*, Ref. 9a, p. 736. (b) N. Phillips *et al.*, in *Progress in low-temperature physics 13*, D. Brewer, Ed. Elsevier, The Netherlands, 1990. (c) A. Junod *et al.*, *Physica C* **159**, 215 (1989).

22. (a) N. Ong *et al.*, *Phys. Rev. B* **35**, 8807 (1987). (b) M. Suzuki, *Phys. Rev. B* **39**, 2312 (1989).

23. D. Morris, M. Mills, and T. Hewston, *J. Chem. Educ.* **64**, 847 (1987).

24. (a) A. Masari *et al.*, *Jpn Appl. Phys. Lett.* **20**, L405 (1987). (b) A. Ramirez *et al.*, *Phys. Rev. B* **35**, 8883 (1987).

25. (a) W. Reichardtdt *et al.*, *Physica C* **162–164**, 464 (1989). (b) M. Rietschel *et al.*, *Physica C* **162–164**, 1705 (1989). (c) W. Reichartdt *et al.*, *Phys. B* **156**, 97 (1989).

26. P. Boeni *et al.*, *Phys. Rev. B* **38**, 185 (1988).

27. H. Morawitz, *Phys. Rev. Lett.* **34**, 1567 (1975).

28. (a) P. Visscher and L. Falicov, *Phys. Rev. B* **3**, 2541 (1971). (b) A. Fetter, *Ann. Phys.* **88**, 1 (1974). (c) P. Hawrylak *et al.*, *Phys. Rev. B* **32**, 4272 (1985).

29. B. Geilikman and V. Kresin, *Sov. Phys.—Doklady* **13**, 1040 (1969).

30. L. Landau and E. Lifshitz, *Quantum mechanics*. Oxford University Press, New York, 1977.

31. V. Ginzburg, *Sov. Phys.—Solid State* **2**, 1834 (1961).

32. G. Deutscher, in Ref. 9a, p. 293.

33. (a) S. Inderhees *et al.*, *Phys. Rev. Lett.* **60**, 1178 (1988). (b) R. Fischer *et al.*, *Physica C* **153**, 1092 (1988). (c) A. Voronel *et al.*, *Physica C* **153**, 1087 (1988).

34. Y. Jean *et al.*, *Phys. Rev. Lett.* **60**, 1069 (1988).

35. (a) R. Ferrell, *Rev. Mod. Phys.* **28**, 308 (1956). (b) J. Carbotte, in *Position solid state physics*, W. Brandt and A. Dupasquier, Eds., p. 32. North-Holland, New York, 1983.

36. R. Benedek and H. Schuttler, *Phys. Rev. B* **41**, 1789 (1990).

37. (a) D. Mattis and J. Bardeen, *Phys. Rev.* **111**, 412 (1958). (b) A. Abrikosov, L. Gor'koy, and I. Khalatnikov, *Sov. Phys.—JETP* **8**, 182 (1959).

38. I. Khalatinikov and A. Abrikosov, *Adv. Phys.* **8**, 45 (1959).

39. J. Chang and D. Scalapino, *Phys. Rev. B* **40**, 4299 (1989).

40. T. Friedman et al., Phys. Rev. B **42**, 6217 (1990).
41. Z. Schlezinger et al., Phys. Rev. Lett. **65**, 801 (1990).
42. (a) H. Haghighi et al., Phys. Rev. Lett. **67**, 382 (1991); Ref. 15, p. 183. (b) M. Peter et al., Europhys. Lett. **18**, 313 (1992).
43. S. Barrett et al., Phys. Rev. B **41**, 6283 (1990).
44. W. Warren et al., Phys. Rev. Lett. **59**, 1860 (1987).
45. H. Piel et al., Physica C **153–155**, 1604 (1988).
46. (a) E. Heyen et al., Phys. Rev. B **43**, 12958 (1991). (b) G. Eliashberg, G. Klimovitch, and A. Rylyakov, J. Supercond. **4**, 393 (1991).
47. (a) A. Kebede et al., Phys. Rev. B **40**, 4453 (1989). (b) J. Torrance et al., Phys. Rev. Lett. **61**, 1127 (1988). (c) J. Neumeier et al., Phys. Rev. Lett. **63**, 2516 (1989). (d) J. Neumeier and B. Maple, Physica C **191**, 158 (1992).
48. S. Anlage et al., Phys. Rev. B **44**, 9164 (1991).
49. W. Pickett, Rev. Mod. Phys. **61**, 433 (1989).
50. A. Fiory et al., Phys. Rev. Lett. **65**, 3441 (1990).
51. H. Piel et al., in High T_c superconductor thin films, L. Correra, Ed. Elsevier, Amsterdam, 1992.
52. (a) M. Gurvitch et al., Phys. Rev. Lett. **63**, 1008 (1989). (b) M. Gurvich and A. Fiory, Phys. Rev. Lett. **59**, 1337 (1987); Ref. 9a, p. 663.
53. N. Phillips et al., Phys. Rev. Lett. **65**, 357 (1990).
54. (a) R. Cava et al., Phys. Rev. B **36**, 5719 (1987); J. Phys. C **165**, 419 (1990). (b) J. Jorgensen et al., Phys. Rev. B **41**, 1863 (1989).
55. H. Paulsen et al., Phys. Rev. Lett. **66**, 465 (1991) Nature **349**, 594 (1991).
56. S. Takahashi and M. Tachiki, Physica C **165–166**, 1067 (1990); **170**, 505 (1990).
57. (a) E. Wolf, Principles of electron tunneling spectroscopy. Oxford University Press, New York, 1985. (b) R. Zeyher, Phys. Rev. B **44**, 10404 (1991).
58. (a) B. Wells et al., Phys. Rev. Lett. **65**, 3056 (1990). (b) X. Wu et al., Phys. Rev. B **43**, 8729 (1992).
59. C. Michel et al., in Advances in Superconductivity II, T. Ishiguro and K. Kajamura, Eds., p. 471. Springer-Verlag, Tokyo, 1990.
60. T. Douglas and E. King, in Experimental thermodynamics, Vol. 1, J. McCullough and D. Scott, Eds. Butterworths, London, 1968.
61. J. Bardeen and M. Stephen, Phys. Rev. A **140**, 1197 (1965).
62. (a) N. Werthamer, E. Helfand, and P. Hohenberg, Phys. Rev. **147**, 295 (1966). (b) U. Welp et al., Phys. Rev. Lett. **62**, 1908 (1989).
63. D. Farrell et al., Phys. Rev. Lett. **64**, 1573 (1990).
64. D. Owen and D. Scalapino, Physica **55**, 691 (1971).
65. B. Geilikman, Sov. Phys.—Solid State **18**, 54 (1976).
66. (a) T. Takahashi et al., Nature **334**, 691 (1988). (b) A. Arko et al., Phys. Rev. B **40**, 2268 (1989). (c) J. Imer et al., Phys. Rev. Lett. **62**, 336 (1989).
67. (a) C. Fowler et al., Phys. Rev. Lett. **68**, 534 (1992). J. Smith et al., Ref. 15, p. 177; J. Smith, in Advances in superconductivity, K. Kjimura and H. Kayakam, Eds., p. 231. Springer, Tokyo, 1991. (b) G. Kido et al., Ref. 67a, p. 237.
68. R. Cava et al., Phys. Rev. Lett. **58**, 4081 (1987).
69. W. Pickett, J. Supercond. **4**, 397 (1991).
70. J. Cohn et al. (unpublished).
71. (a) A. Jezowski et al., Phys. Lett. A **122**, 433 (1987). (b) C. Uher, J. Supercond. **3**, 337 (1990).

72. B. Geilikman and V. Kresin, *Kinetic and non-steady effects in superconductors.* Wiley, New York, 1974.

73. (a) S. Barrett *et al., Phys. Rev. Lett.* **66**, 108 (1991). (b) U. Welp *et al.* (unpublished).

74. T. Hasigawa, H. Ikuta, and K. Kitazawa, in *Physical properties of high temperature superconductors III*, D. Ginsberg, Ed., p. 525. World, Singapore, 1992.

75. (a) F. Borsa *et al., Phys. Rev. Lett.* **68**, 698 (1992). (b) J. Martindalle *et al., Phys. Rev. Lett.* **68**, 702 (1992).

76. S. Vedeneev *et al.* (preprint).

77. N. Hass *et al., J. Supercond.* **5**, 191 (1992).

78. J. Valles *et al., Phys. Rev. B* **44**, 11986 (1991).

79. (a) Z. Schlesinger *et al., Phys. Rev. Lett.* **55**, 801 (1990). (b) Z. Schlesinger *et al.,* Ref. 15, p. 147.

80. *J. Supercond.* **3**, No. 3 (1990).

81. G. T. Lee, J. R. Collman, and W. A. Little, *J. Supercond.* **3**, 197 (1990).

82. N. Bulut and D. Scalapino, *Phys. Rev. Lett.* **68**, 706 (1992).

83. Q. Huang *et al., Nature* **347**, 369 (1990).

84. (a) J. Torrance *et al., Phys. Rev. Lett.* **61**, 1127 (1988). (b) H. Takagi *et al., Phys. Rev. B* **40**, 2254 (1989).

85. Y. Takmura *et al., Phys. Rev. B* **38**, 7156 (1988).

86. J. Hulm *et al., Proc. LT-X.* VINITI, Moscow, 1967.

87. V. Kresin, *J. Low Temp. Phys.* **5**, 565 (1971).

88. (a) G. Yu *et al., Solid State Commun.* **72**, 345 (1989). (b) V. Kudinov *et al., Phys. Lett.,* **A151**, 358 (1990). (c) G. Nieva *et al., Appl. Phys. Lett.* **60**, 2159 (1992).

89. (a) J. Mannhart *et al., Z. Phys. B* **83**, 307 (1991). (b) J. Mannhart (unpublished).

90. H. Takagi *et al., Phys. Rev. Lett.* **69**, 2975 (1992).

91. Y. Uemura, *Phys. Rev. Lett.* **62**, 2317 (1989).

92. T. Schneider and M. Keller, *Phys. Rev. Lett.* (in press).

93. (a) B. Batlogg *et al.,* Ref. 9a, p. 653. (b) M. L. Cohen, Ref. 9a, p. 733.

94. G. Gladstone, M. Jensen, and J. Schrieffer, in *Superconductivity,* R. Parks, Ed., p. 665. Marcel Dekker, New York, 1969.

95. M. Crawford *et al., Phys. Rev. B* **41**, 282 (1990).

96. (a) Franck, *Physica C* **172**, 90 (1990); Ref. 15, p. 411. (b) H. Bornemann and D. Morris, *Physica C* **182**, 132 (1991).

97. (a) J. Labbe and J. Bok, *Europhys. Lett.* **3**, 1225 (1987). (b) I. Dzyaloshinskii, *Sov. Phys.—JETP* **66**, 848 (1987). (c) C. Tsuei *Phys. Rev. Lett.* **65**, 2724 (1990). (d) C. Kane *et al.,* Ref. 15, p. 493.

98. V. Kresin and S. Wolf (unpublished).

99. (a) R. Cohen, W. Pickett, and H. Krakauer, *Phys. Rev. Lett.* **64**, 2575 (1990). (b) H. Krakauer, W. Pickett, and R. Cohen, *Phys. Rev. B* **47**, 1002 (1993).

100. (a) R. Greene and P. Chaikin, *Phys. Rev. B* **126**, 431 (1984). (b) T. Ishigura and K. Yamaji, *Organic Superconductors.* Springer-Verlag, Heidelberg, 1990. (c) *Organic Superconductivity,* V. Kresin and W. A. Little, Eds. Plenum Press, New York, 1991.

101. S. Wolf and V. Z. Kresin, Ref. 100c, p. 31.

102. K. Oshima *et al., Phys. Rev. B* **38**, 938 (1988).

103. T. Osada *et al.,* Technical Report of ISSP, Serial A #2211 (1989).

104. Y. Uemura, Ref. 100c, p. 23.

105. T. Mori *et al.*, *Bull. Chem. Soc. Japan* **57**, 627 (1986).
106. K. E. Gray, M. E. Hawley, and E. R. Moog, Ref. 9a, p. 611.
107. D. Jerome and F. Creuzet, Ref. 9a, pp. 103–135.
108. M. Ichimura *et al.*, *Phys. Rev. B* **41**, 6387 (1990).
109. A. Hebard *et al.*, *Nature* **350**, 600 (1991).
110. M. Rosseinsky *et al.*, *Phys. Rev. Lett.* **66**, 2830 (1991).
111. K. Holczer *et al.*, *Science* **252**, 1154 (1991).
112. S. Kelly *et al.*, *Nature* **352**, 222 (1991).
113. H. Kroto *et al.*, *Nature* **318**, 162 (1986).
114. W. Kruetschmer *et al.*, *Nature* **347**, 354 (1990).
115. R. Haddon *et al.*, *Nature* **350**, 320 (1991).
116. A. Hebard, *Physics Today* **45**, 26 (1992).
117. A. P. Ramirez *et al.*, *Phys. Rev. Lett.* **68**, 1058 (1992).
118. M. Schluter *et al.*, *Phys. Rev. Lett.* **68**, 526 (1992).
119. Y. Marumara, MRS Meeting (Abstracts, Boston, 1992). MRS, Pittsburg, 1992.
120. V. Z. Kresin, *Phys. Rev. B* **46**, 14883 (1992).
121. R. Tycko *et al.*, *Phys. Rev. Lett.* **68**, 1912 (1992).
122. P. Allen and R. Dynes, *Phys. Rev.* **312**, 905 (1975).
123. Z. Zhang, C. Chen, and C. Lieber, *Science* **254**, 1619 (1991).
124. V. Z. Kresin, Ref. 100c, p. 285; V. Kresin, V. Litovchenko, and A. Panasenko, *J. Chem. Pnys.* **63**, 3613 (1975).
125. M. Klein, *Physica C* **162**, 1701 (1989).

INDEX